Water Resources Development and Management

Indexed by Scopus

Each book of this multidisciplinary series covers a critical or emerging water issue. Authors and contributors are leading experts of international repute. The readers of the series will be professionals from different disciplines and development sectors from different parts of the world. They will include civil engineers, economists, geographers, geoscientists, sociologists, lawyers, environmental scientists and biologists. The books will be of direct interest to universities, research institutions, private and public sector institutions, international organisations and NGOs. In addition, all the books will be standard reference books for the water and the associated resource sectors.

More information about this series at http://www.springer.com/series/7009

Asit K. Biswas · Pawan K. Sachdeva ·
Cecilia Tortajada

Phnom Penh Water Story

Remarkable Transformation of an Urban
Water Utility

 Springer

Asit K. Biswas
University of Glasgow
Glasgow, UK

Pawan K. Sachdeva
Water Management International
Private Ltd.
Singapore, Singapore

Cecilia Tortajada
Institute of Water Policy
Lee Kuan Yew School of Public Policy
National University of Singapore
Singapore, Singapore

ISSN 1614-810X   ISSN 2198-316X   (electronic)
Water Resources Development and Management
ISBN 978-981-33-4067-1   ISBN 978-981-33-4065-7   (eBook)
https://doi.org/10.1007/978-981-33-4065-7

This Springer imprint is published by the registered company Springer Nature Singapore Pte Ltd.
The registered company address is: 152 Beach Road, #21-01/04 Gateway East, Singapore 189721,
Singapore

Foreword

When history looks back at 2020, it will be remembered as a point in time where the entire world had to battle against novel coronavirus. This pandemic has reminded all of us having access to safe water to wash hands is vital in protecting our own lives and those of our loved ones. The people of Phnom Penh, where this story takes place, do not have to worry about whether they can access clean water to wash their hands or have enough water to support their economic activities. This is because despite being a city in a developing country, Phnom Penh has miraculously enabled their citizens to have 24 h access to safe, drinkable water. Phnom Penh's incredible water supply service was created in just over the course of a decade and has continued to revitalize the country whose infrastructure and resources were previously destroyed from long-lasting conflict and turmoil. In 1993, before the start of this transformation, only 20 per cent of the population in Phnom Penh had access to tap water. Additionally, only the limited number of staff had the professional skills in water management and staff morale was not always high. On top of this, the city's water tariffs accounted for only a third of the revenue needed to distribute clean water to its residents. This book, therefore, provides a complete picture of how Phnom Penh, despite such challenges, was able to successfully revitalize its water supply service and overcome its many adversities.

The Japan International Cooperation Agency (JICA), as a long-standing partner of the Phnom Penh Water Supply Authority (PPWSA), has had the privilege of being involved in this miraculous story of revitalizing Phnom Penh. While this research is independent from JICA, as a development partner that promotes social and economic development across the globe, we seek to recognize three fundamental elements that we believe underpin Phnom Penh's continued success.

First, we credit the unwavering dedication of the staff at PPWSA, led by its General Director, Ek Sonn Chan, who served as the primary architect in making this miracle possible. The PPWSA committed itself to taking on the enormous task of establishing a reliable water utility that ensures safe water supplies for all of Phnom Penh's residents. This outstanding leadership can be largely credited for the increased accountability and transparency exhibited across the city's water utility. While development partners did play a supportive role in this endeavour, the success of this story is truly a result of the strong will and determination of the PPWSA. Without the dedication of local partners, external development cooperation itself cannot be wholly effective or sustainable. This book, therefore, emphasizes how the PPWSA's remarkable improvements over a short period of time were achieved primarily due to the leadership of the Cambodian people, including by the water utility's dedicated managers.

The second element that underpins this success story is the PPWSA's emphasis on improving and expanding its services to keep pace with the city's urban development and rising living standards. In order to achieve the Sustainable Development Goals (SDGs) and ensure water and sanitation access for all, water utilities must become financially sound and reinvest in their own facilities to expand and adapt their services to meet the needs of society. Under the strong leadership at PPWSA, the water utility was able to incentivize its employees to transform itself to do just that. The PPWSA initially took advantage of Japan's grant aid to rehabilitate its facilities, and then it worked to become financially self-sufficient and eventually was able to repay its loans. Afterwards, the PPWSA succeeded in going public, becoming a partially publicly traded and owned entity. Throughout this process, the water utility continued to improve its tariff collection rate, reduce water leaks and provide professional development for their staff—serving as a good model to replicate in other developing countries to achieve the SDGs.

The third major element that makes Phnom Penh's story a remarkable success was the ability of the PPWSA to share its knowledge and replicate its experience, not only across Cambodia, but also beyond its borders. After becoming Cambodia's Secretary of State for Water Supplies, Ek Sonn Chan, the former General Director of the PPWSA, prioritized efforts to dispatch several of his former staff to lead public water utilities in eight of Cambodia's major provincial cities. There, the former PPWSA staff successfully navigated in turning around the water utilities in their cities. Once Cambodia achieved this miraculous transformation, Ek Sonn Chan then worked to share the PPWSA's experiences with other developing countries through international conferences and forums and by conducting training programmes with development partners. Through these discussions, Cambodia's experience was heard by numerous high-level government officials, including the Minister of Water Resources and Irrigation of South Sudan. The Minister was so impressed by the PPWSA's ability to transform Cambodia's war-torn water supply system that he decided to draft a revitalization plan for South Sudan emulating the PPWSA experience. Additionally, the PPWSA also greatly inspired Myanmar's Yangon City Development Committee. Immediately after participating in a training course led by the PPWSA, the Yangon City Development Committee decided to build its own training centre for its water supply service after realizing the importance of investing more in human resource development. In this sense, South Sudan and Myanmar came to understand that their water utilities could also benefit from learning about Phnom Penh's experience. Similarly, this book has used Phnom Penh's story to provide seven key learnings and ten recommendations to provide guidance for leaders of water utilities around the world.

While the PPWSA truly spurred a miraculous transformation within Phnom Penh, the water utility is now facing new challenges in sustaining its success. Phnom Penh is developing faster than ever, rapidly constructing large office buildings, commercial complexes and factories. This situation has led to an enormous increase in water demand, requiring the PPWSA to expand its facilities more systematically. To finance these endeavours, the PPWSA will have to navigate fair water tariffs for those who have economically benefited from such development, while also protecting the poor. For its part, JICA will continue to cooperate with the city in expanding its water treatment plants and reviewing its master plan.

Aside from the city's rapid development, the PPWSA is also concerned about preserving its institutional knowledge. The staffers who were central in Phnom Penh's miraculous transformation since 1993 are now fast approaching retirement. Therefore, the PPWSA is working to invest more in human resource development to avoid institutional memory loss. They need to cultivate future leaders that are mission-driven and have strong problem-solving skills. In support of these endeavours, JICA, under its Development Studies Program (JICA-DSP), offers opportunities for PPWSA staff to take postgraduate courses at Japanese universities. Given its impressive history of solving very complex problems, we are confident that the PPWSA will be able to similarly overcome these new challenges.

Lastly, I would like to congratulate Asit K. Biswas, Pawan K. Sachdeva and Cecilia Tortajada, for their phenomenal job at sharing Phnom Penh's success story in a way that is easy to understand for all audiences. I wholeheartedly recommend this book to leaders from around the world, as well as to future leaders in public policy and development cooperation. This book not only details the journey of Phnom Penh's success, but also presents a universal framework on how to best analyse the capacity of water utilities from four domains, and it provides policy-makers and water utility mangers with ten clear recommendations on how to improve safe water access.

By sharing the *Phnom Penh Water Story*, we hope many cities around the world will be able to replicate these outcomes and help realize a world where people have universal access to safe water.

Tokyo, Japan

Kitaoka Shinichi
President
Japan International Cooperation
Agency (JICA)

Preface

Provision of clean water to an increasing number of people in the urban areas of developing countries has been a major problem for decades. The problem started to worsen steadily in most urban centres, especially after the post-1960 period when the rate of urbanization increased steadily. The number of people to whom the water had to be provided and the high rate of urbanization simply overwhelmed the limited administrative, management, financial and institutional capacities of nearly all cities of the developing world.

In the case of Phnom Penh, the situations in the 1980s and the early 1990s were very different compared to other cities of the world. This is because of the extraordinary conditions Phnom Penh and Cambodia had to face because of the brutal Khmer Rouge rule. Its forces took over Phnom Penh and then the rest of Cambodia as a whole in 1975. Its leader, Pol Pot, declared the nation will start at "year zero", proceeded to isolate Cambodia from the rest of the world, started forcibly to empty the cities by moving urbanites to rural areas and abolished currency, private property and religion. The Khmer Rouge regime was eventually overthrown by Vietnamese forces in 1979.

Within this brief 4-year period of Khmer Rouge rule, they played havoc with Cambodia's people, economy and infrastructure. For the first time in modern history, all Cambodian cities were depopulated and their infrastructure and management, including in the water sector, were basically destroyed.

The decade of the 1980s, mostly because of international geopolitics, was not the best decade for Cambodia in terms of its potential social and economic development. For nearly all purposes, it was a lost decade for the country.

Because of these unfortunate happenings, not surprisingly, Phnom Penh's water supply was in a perilous state during the 1980s and the early 1990s. Throughout this period, only a limited number of households received water of dubious quality for only very few hours each day. The utility had limited technical, administrative and managerial capacities. Corruption was widespread. For all practical purposes, the utility was bankrupt. All these malaises were well known. Given the social, economic and political turmoils, the water utility and all other public services were in poor shape for nearly three decades.

It was thus somewhat of a surprise, when in early 2005, a water professional specializing in Asian water utilities asked me what I thought of the remarkable transformation of Phnom Penh Water Supply Authority (PPWSA), under its Director General Ek Sonn Chan. I told him PPWSA was not in my radar, but given the country's conditions during the early 1990s, the chances of it making a remarkable transformation, especially given the then prevailing social and economic conditions, must have been realistically close to zero.

The following year, in 2006, I suddenly received a totally unexpected invitation to visit Phnom Penh to address Prime Minister Hun Sen, his entire cabinet and all the provincial governors, on how Cambodia's water management can be substantially improved, for all types of uses, and thereafter have a dialogue with his Ministers and their senior officials. I accepted this invitation with alacrity. A side benefit I expected would be to visit PPWSA and check out if my friend's view that it had undergone "remarkable transformation" was accurate.

During this 2 h meeting which Prime Minister Hun Sen chaired, in his closing address he formally invited me to visit PPWSA and then give his government my views of how it had progressed, and then advise how it can be improved. It was evident that Hun Sen was very proud of PPWSA's achievements.

Director General of PPWSA, Ek Sonn Chan, was also present at this high-level meeting. Shortly thereafter, he took me to visit PPWSA headquarters. This was the beginning of my association with Ek Sonn Chan and his senior staff. What I saw was that contrary to all expectations, PPWSA had achieved, under very challenging social and economic conditions, a truly remarkable transformation. Equally impressive was the humility of Ek Sonn Chan and his senior staff in spite of what nearly every water expert might have considered to be near impossible to accomplish under such demanding circumstances.

Since then, myself, Prof. Cecilia Tortajada and Pawan K. Sachdeva, my partner at Water Management International Pte Ltd. of Singapore, have visited Phnom Penh several times. We received all the data we needed to analyse PPWSA comprehensively. We also conducted several independent interviews with users of water, domestic and industrial, rich and poor. Pawan crunched all the numbers to evaluate its performance over the past two decades, and Cecilia and I worked on assessing the enabling environment which made this near miracle possible. This book, *Phnom Penh Water Story*, is the result.

One of the issues we were interested in was how PPWSA has fared after leadership change, under a new Director General, Sim Sitha. Our view is that PPWSA is marching ahead successfully, even though many of the challenges PPWSA has faced under his leadership have been somewhat different than earlier.

Phnom Penh Water Story is truly an uplifting one. To our knowledge, no other city in any other developing country has achieved so much, even under such very difficult conditions. This confirms my strong personal view that the world is not facing a water crisis because of physical lack of this resource. However, it is indeed facing a major water management crisis, the magnitude and extent of which no other earlier generation had to face. The two crises are very different, and their solutions are also very different.

The question that now should be asked, and answered, is if Phnom Penh, with all challenges and constraints, could provide all its citizens clean water, 24 hours a day, and seven days a week, which can be drunk straight from the taps without any health concerns, why other cities of developing countries like Delhi, Dhaka, Lagos, Cairo or Buenos Aires cannot do the same? All these cities have significantly more technical and administrative expertise and financial wherewithal than Phnom Penh when PPWSA started its remarkable journey in 1993. Phnom Penh actually faces significantly harder and more difficult and complex challenges compared to all its counterparts in nearly all developing countries. If Phnom Penh with all the con straints it faced in 1993 did it within one decade, there is absolutely no reason why other cities of developing countries, which are in a much better position managerially, technically and financially, should not be able to perform the same feat.

This book would not have been possible without the unstinted help and support of Ek Sonn Chan and Sim Sitha, and all other PPWSA's staff members, senior and junior, who made PPWSA's remarkable water journey possible. I would especially like to thank Dr. Chea Visoth and Mr. Samreth Sovithiea for their friendship and support ever since my first visit in 2006. Without their help, it would not have been possible to write this definitive book. They provided us with all the information and data we needed promptly and then checked and rechecked many aspects of our analyses. However, this book reflects the views of the authors only. Based on our own experiences, it is one of the very few water utilities anywhere in the world that is ensuring complete data accessibility and transparency.

Finally, we are most grateful to Ms. Thania Gómez of the Third World Centre for Water Management, for her continuous and unstinted help to transform the entire manuscript, including figures and tables, to Springer's format. She has played an essential role in finalizing the manuscript.

Singapore/Glasgow, UK   Asit K. Biswas
September 2020   Distinguished Visiting Professor
                                                                              University of Glasgow
                                                                                                         UK

                                                                                               Director
                                           Water Management International Private Ltd.
                                                                                          Singapore

                                                                               Chief Executive
                                           Third World Centre for Water Management
                                                                                            Mexico

Contents

Abbreviations

ADB	Asian Development Bank
AFD	French Development Agency
CNRP	Cambodia National Rescue Party
CPP	Cambodian People's Party
GDP	Gross Domestic Product
JICA	Japan International Cooperation Agency
KHR	Cambodian Riel
MIH	Ministry of Industry and Handicrafts, now Ministry of Industry, Science, Technology and Innovation (MISTI)
MPWT	Ministry of Public Works and Transportation
MRD	Ministry of Rural Development
NCDD	National Committee for Democratic Development
NRW	Non-Revenue Water
PPM	Phnom Penh Municipality
PPWSA	Phnom Penh Water Supply Authority
PUB	Public Utilities Board, Singapore
SWSA	Sihanoukville Water Supply Authority
UFW	Unaccounted for Water
USD	US Dollar
WB	World Bank
WTP	Water Treatment Plant

List of Figures

List of Tables

Chapter 1
About This Book

1.1 Book Outline

Phnom Penh is the capital city and is also the political and the economic centre of the Kingdom of Cambodia. Ravaged by internal conflicts and social and political turbulences for decades, Phnom Penh's drinking water infrastructure, as well as the city's overall management of all other services and infrastructure, were in shambles in 1993. Any sane person, resident or visitor to the city, in 1993, would have been utterly disappointed with the domestic water supply situations, both in terms of quantities available and their qualities. The water utility was non-functional, management was dismal, all water infrastructures were dilapidated and in urgent need of rehabilitation, technical capacity was non-existent and the utility was bankrupt for all practical purposes. Phnom Penh's water supply system at that time was one of the very worst in any capital of any developing country. Fast forward to 2008, after this dismal situation, the city's residents were receiving a continuous water supply of good quality that could be drunk straight from the tap without any health concerns. The water utility was profitable, financially sustainable, and management and technical capacities were vastly improved. In only one decade, Phnom Penh's water utility improved from one of the worst in the developing world to be one of the very best.

This book is a comprehensive documentation and objective analysis of this remarkable transformation of the Phnom Penh Water Supply Authority (PPWSA), the institution that has been responsible for providing drinking water supply services to the city, from an almost failed and bankrupt institution in 1993 to be one of the very few successful and financially viable water utility in any city of any developing country of the world.

In 1993, the gross domestic product (GDP) per capita for Cambodia was only USD 254 per year, that is, little over USD 20 per month. Given the abject economic conditions, widespread poverty, malfunctioning institutions, and social and political challenges of the early 1990s, PPWSA's institutional transformation and its overall performance to provide all the inhabitants of Phnom Penh clean, reliable and continuous water supply is a beacon of inspiration and hope to other water utilities of

© The Author(s), under exclusive license to Springer Nature Singapore Pte Ltd. 2021
A. K. Biswas et al., *Phnom Penh Water Story*, Water Resources Development
and Management, https://doi.org/10.1007/978-981-33-4065-7_1

the world, both in developing and developed countries. It is truly a remarkable and uplifting story, one that has not been told objectively and comprehensively ever before. This book provides a thorough and detailed analysis of the performance of PPWSA during the 1993–2017 period, and the enabling conditions that made this remarkable transformation possible.

The book also produces a framework for conducting a strategic and implementable analysis for any urban water utility of the developing world. The framework proposed divides the analysis of urban water utility into four distinct and interrelated domains of analysis. These are physical, operational, financial and institutional domains. PPWSA's performance during the period of 1993–2017 has been analysed using this new proposed four-domain framework.

This book has two primary goals. The first goal is to share the PPWSA's successful turnaround story in little more than one decade as an inspirational and aspirational achievement for urban water utilities of other developing country cities to follow and learn from. It can motivate other water utilities of the developing world that provision of 24 × 7 clean drinking water is not only possible but also can be done successfully in about a decade. The story is remarkable especially when one considers the social, economic and political challenges that prevailed in Cambodia during the time when its radical transformation in water supply services took place. The fact that PPWSA succeeded under such demanding conditions indicates that the urban water utilities of other developing countries can also flourish, provided they approach possible solutions to their problems systematically, logically and on a long-term basis.

The second goal is to introduce a strategic framework for analysing any urban water utility of any developing country cities that could ensure its success and long-term sustainability. Unfortunately, such a logical and implementable framework, for the most part, has not been available thus far. This framework should enable managers of water utilities of any developing country to identify, implement and then resolve the constraints they are facing to have a fully functional and financially viable water supply system.

1.2 Readership of the Book

The primary audience of this book are policymakers, managers and water professionals of the developing world who are associated with urban water planning and management. They can read the inspiring story of Phnom Penh Water Supply Authority and draw lessons from its success which can be implemented in their respective water utilities, after appropriate modifications to suit their specific local conditions. They can also further benefit from the use of the four-domain framework outlined in this book. They could analyse the performances of their own water utilities in the context of this framework and benchmark their institutions' achievements with other water utilities to see how well they may be discharging their tasks.

Urbanisation has been a megatrend in developing countries for decades (Varis 2018). With increasing population growth, urbanisation and industrialisation, lack of reliable availability of adequate clean water has rapidly emerged as one of the important limiting factors for further economic growth and social development of urban areas. Accordingly, anyone interested in provision of water services to urban settlements is likely to find this book useful to ensure provision of clean water to their cities on a regular and continuous basis.

The book is also intended for the policymakers responsible for formulating and implementing evidence-based policies for ensuring reliable urban water services are available. The methodology outlined can be used as a useful tool to evaluate the existing status of urban water services, and also assess the current gaps in the four domains outlined. The problems of the urban water services sector in developing countries have been well documented in the literature. This book will give the policymakers and managers hope that, like Phnom Penh, their cities should also be able to solve the problems associated with the provision of clean water on a 24 × 7 basis. It will further provide practical information on the sequencing of solutions that has been achieved by a successful institution like PPWSA.

This book is meant for academics, research scientists, students interested in the provision of reliable and safe urban water services, as well as for professionals and policymakers associated with urban water supply systems. It provides the case study of PPWSA as an example of how the problems in a real-world situation were successfully solved in one decade. It also provides full documentation and an up-to-date authoritative analysis of PPWSA's performance through the lens of the proposed four-domain framework.

1.3 Background to the Book

All the three authors have long been associated with PPWSA as mentors and honorary advisors. The involvement started in 2006, when the senior author, Prof. Asit K. Biswas, received an invitation from the Prime Minister Hun Sen of Cambodia, to address a meeting chaired by the Prime Minister, with all the Ministers and Governors of the Provinces and senior public officials of the country present. This invitation was extended by the Prime Minister through the Asian Development Bank, shortly after Prof. Biswas received the Stockholm Water Prize, considered to be the equivalent of the Nobel Prize in the area of water, in August 2006. The objective of this meeting was to assess the current status of access to clean water supply in Phnom Penh and other urban areas of Cambodia and how the situations in all Cambodian cities could be significantly improved.

During this very specially convened meeting of the Ministers and all the Governors, Prime Minister Hun Sen formally requested Prof. Biswas if he could visit PPWSA and advise how the performance of PPWSA could be further enhanced. Following this high-level meeting, Prof. Biswas had meetings with several Ministers on different aspects of water management in Cambodia, including the Mekong. These focused primarily on the many challenges Cambodia was facing in the water sector, and how they could be realistically solved within the then prevailing social and economic conditions of the country.

The then Director General of PPWSA, Ek Sonn Chan, took Prof. Biswas for discussions with the senior officials of the utility. He then subsequently visited its headquarters and water supply systems.

Four factors during this first visit impressed Prof. Biswas. First, everyone in the institution, starting from the Director General, was drinking water straight from the tap without any hesitation. There was no sign of any bottled water in any of the offices of the senior officers on the utility. This indicated that staff members were confident with the quality of water they were producing and distributing, so much so that they had no hesitation in drinking it. Second, the water treatment works and water quality laboratory were both scrupulously clean, even in better conditions than in many cities of developed countries. During the visit to the water treatment plant, there was a small pebble on one of the dividing walls. The Director General nonchalantly picked it up and put it in his trouser pocket. This indicated that senior officers were interested in all big and small issues. Third, compared to the commotion and hubbub outside, PPWSA was peaceful and staff were busily working in properly organised offices, within a good landscaped area and a well-maintained garden. This was an indication that the staff, from top to bottom levels were dedicated and proud of their work. Finally, the washrooms of normal staff members were scrupulously clean and very well maintained. All these conditions can rarely be seen in nearly all water utilities of the developing world.

Fourth was an anecdote Prof. Biswas heard from a very senior Cambodian official. Shortly after Ek Sonn Chan was appointed, Cambodian government auditors arrived in PPWSA to carry out their audit. They offered the Director General to do a light audit if they could receive some financial "incentives." The Director General promptly rejected the offer. Instead he requested the auditors to conduct a tough and thorough audit as they could do so that he could find out what were the shortcomings and then promptly rectify them. The surprised auditors did a quick audit and left for other likely more lucrative assignments! This is likely to be a true story since during Prof. Biswas' first meeting with all the senior staff of PPWSA, Ek Sonn Chan made an "humble request" to Prof. Biswas. He first confirmed PPWSA would promptly provide all the information Prof. Biswas wanted. Following the visit and analyses of the data, if he could kindly inform the Director General what they have been doing could be further improved. If so, how? Also, should they be doing something extra which could improve the services they were providing to the public.

Impressed by these factors, and his discussions with the Director General and all his senior staff members, Prof. Biswas accepted the invitation to be a mentor to the institution and have been advising the institution ever since.

From 2007 onwards, both Biswas and Cecilia Tortajada made several visits to Phnom Penh and had numerous discussions with all PPWSA senior officials, senior civil servants from different Government agencies and aid organisations like Japan International Cooperation Agency, Asian Development Bank, World Bank, other development agencies, national and international NGOs in water and other development sectors, and educational institutions.

These visits, analyses of data collected, and interviews with large and small-scale water users, resulted in a series of authoritative analyses which brought the success of the Phnom Penh water story to global attention (Biswas 2010; Biswas and Tortajada 2009, 2010; Tortajada and Biswas 2019).

The work of Biswas and Tortajada considered PPWSA's performance until 2009. Impressed by the remarkable transformation of PPWSA, from being one of the basket cases of water utilities in 1993 to be one of the best in any city of the developing world within a short period of about a decade, Biswas and Tortajada decided to bring this extraordinary performance to the attention of the world. Accordingly, they nominated PPWSA, under the leadership of Ek Sonn Chan, for the prestigious Water Industry Award of the Stockholm World Water Week, in 2010. The nomination, not surprisingly, was successful. In awarding the Prize, the jury noted: "The PPWSA has a strong commitment to social and environmental responsibility. It has shown the developing world as a whole that large cities can expect continuous access to clean water. It stands as a role model for those committed to improving their businesses practices and increasing their level of services to customers."

During a forceful and emotional acceptance speech, Director General Ek Sonn Chan first appreciated the work of Biswas and Tortajada as mentors and advisors, and then went to acknowledge his debt to them for their objective analyses of the performance of PPWSA and publishing them in major international journals which brought the institution to the attention of the world.

Pawan Sachdeva joined Biswas and Tortajada to further study PPWSA in 2016 and also to collect latest available information, analyse all the additional new information and to develop a framework by which water utilities of the developing world can be effectively analysed. He made several trips to Phnom Penh to discuss with Sim Sitha, current Director General of PPWSA and all its other senior staff, as well as appropriate government agencies associated with PPWSA, including Ministry of Industry and Handicraft (MIH), Ministry of Public Works and Transport (MPWT), and Phnom Penh offices of Japan International Cooperation Agency (JICA), and Asian Development Bank (ADB). He also has had numerous discussions with Ek Sonn Chan, the previous Director General of PPWSA, under whom all the earlier developments had taken place. Sachdeva also conducted 20 primary interviews with selected residents of Phnom Penh to obtain their views on the extent and qualities of urban services they are receiving, including access to water, its availability and quality. Biswas and Tortajada have also made additional trips to Phnom Penh to better understand and appreciate their post-2009 progress.

These long associations of the three authors with PPWSA will provide an accurate, complete and objective picture of the performance of PPWSA until at least 2017.

1.4 Book Structure

The book is divided into several chapters. Chapter 2 provides a brief background of social, economic and political conditions of Cambodia so that the readers can fully understand and appreciate the challenging social, political, economic and institutional conditions within which PPWSA had to work with. Chapter 3 provides a comprehensive documentation and analysis of Phnom Penh Water Supply Authority's performance from 1993 to at least 2017, but in most instances up to 2019. This chapter also considers the political, administrative and economic conditions of Cambodia, as well as Phnom Penh, during this period. It introduces a framework for analysing an urban water utility, especially in the developing world, through four interrelated and interacting domains: physical, operational, financial and institutional. This framework is subsequently used to analyse the performance of PPWSA. The remaining chapters assess the challenges that PPWSA is likely to face in the coming years and discuss the lessons that water utilities from cities of the developing world can learn, and possibly use, from the successful transformation of PPWSA. The last chapter provides a step wise-guide for analysing an urban water utility.

References

Biswas AK (2010) Water for a thirsty urban world. Brown J World Affairs 17(1):147–166. https://www.jstor.org/stable/24590763

Biswas AK, Tortajada C (2009) Water supply of Phnom Penh: a most remarkable transformation. Research Report, Third World Centre for Water Management, Mexico

Biswas AK, Tortajada C (2010) Water supply of Phnom Penh: an example of good governance. Int J Water Resour Dev 26(2):157–172. https://doi.org/10.1080/07900621003768859

Tortajada C, Biswas AK (2019) Objective case studies of successful urban water management. Int J Water Resour Dev 35(4):547–550. https://doi.org/10.1080/07900627.2019.1613766

Varis O (2018) Population megatrends and water management. In: Biswas AK, Tortajada C, Rohner P (eds) Assessing global water megatrends. Springer, Singapore, pp 41–59

Chapter 2
Understanding Cambodia

2.1 Cambodia

2.1.1 Political Background

The Khmer empire was the predecessor state to modern Cambodia. Angkor was its capital city till the early fifteenth century. Phnom Penh became the royal capital for 73 years, from 1432 to 1505. However, at that time, it was known as "Chaktomuk" (Four Faces), because of its location next to the four-branched confluence of the rivers Mekong, Tonlé Sap and Bassac. Phnom Penh derives its current name from the legend Daun Penh (Grandmother Penh). It literally means Penh's hill (in Khmer language *Phnom* means hill).

From 1866, Phnom Penh became the permanent seat and the capital of Cambodia under the reign of King Norodom, who ruled Cambodia from 1860 to 1904. Cambodia was integrated into the French Indochina Union in 1887, along with other French Colonies and Protectorates in Vietnam. Cambodia gained its independence from the French on 9 November 1953.

Norodom Sihanouk became the Prime Minister after the Cambodian general election of 1955. Norodom Sihanouk was the King of Cambodia from 1941 to 1955. He abdicated his throne in favour of his father Norodom Suramarit, in 1955, to fight the 1955 general election. After the death of Norodom Suramarit in 1960, Norodom Sihanouk made a constitutional amendment and became the head of state from 1960 to 1970.

General Lon Nol became the commander-in-chief of the armed forces in 1960, and was a trusted aide of Norodom Sihanouk. In 1970, General Lon Nol staged a coup and ousted his patron Prince Norodom Sihanouk, and forced him to exile into China. General Lon Nol became the Prime Minister of Cambodia during 1966 to 1967, and then again during 1969–1972. From 1972 to 1975, General Lon Nol was the President of the Khmer Republic.

General Lon Nol's reign did not have full control of the entire Cambodia. Khmer Rouge, the followers of the Communist Party of Kampuchea, formed in 1968, was

© The Author(s), under exclusive license to Springer Nature Singapore Pte Ltd. 2021
A. K. Biswas et al., *Phnom Penh Water Story*, Water Resources Development
and Management, https://doi.org/10.1007/978-981-33-4065-7_2

in control of some parts of Cambodia. In 1975, the Khmer Rouge (meaning "Red Khmers") emerged victorious after the Cambodian Civil War. They overthrew the military dictatorship of General Lon Nol and installed their own government. They named the country as Democratic Kampuchea.

Khmer Rouge attempted to turn Cambodia into a rural and classless society by depopulating urban centres and forcing city dwellers to work in rural agricultural communes. Shortly after assuming power, Khmer Rouge forced nearly two million people from Phnom Penh and other urban areas into the countryside for agricultural work. They wanted to transform Cambodia into a country where there were no rich or poor people, and no exploitation of the people. They abolished money, free markets, schools and colleges, private property, and religious practices. There was no public or private transportation. Shortly, after seizing power, they arrested and killed thousands of civil servants, soldiers and officers. Over the following three years, they executed hundreds of thousands of intellectuals, urban residents and minorities whom they considered to be traitors.

Various studies have estimated that the total death tolls during the Khmer Rouge regime ranged between 740,000 and 3,000,000. The most common estimates range from 1.4 to 2.2 million deaths, which were primarily because of executions and due to the policies they pursued which resulted in widespread malnutrition, starvation, diseases and then death.

Vietnamese armed forces finally captured Phnom Penh on 7 January 1979 and formed a new government. The name of the country was changed to People's Republic of Kampuchea. Khmer Rouge was forced to retreat to western Cambodia and continued to control certain specific areas near the border with Thailand during the following decade. Because of geopolitical reasons and the fact that Vietnam was believed to be in the Soviet Camp, the United States and China continued to support the government in exile that comprised the Khmer Rouge and some other factions even though most of the country was under the new government.

In 1985, Vietnam announced that it would withdraw its forces by 1990. It did so earlier by 1989. The name of the country was again changed, in 1989, from People's Republic of Kampuchea to the State of Cambodia. After protracted decade-long negotiations, on October 23, 1991, the Agreements on a Comprehensive Political Settlement of the Cambodia Conflict were signed between Cambodia and 18 other nations under the aegis of the United Nations.

History of modern Cambodia can be considered to have started from about 1991. There have been general elections in Cambodia every five years since 1993. The last general election was held in 2018. The Cambodian People's Party (CPP) has been the ruling political party of Cambodia for decades.

2.1.2 Administrative Arrangements

Cambodia is a constitutional monarchy. The constitution of Cambodia vests exclusive power in the National Assembly and the governance and general administration are conducted through ministries at the national level and the provincial administrations (IWMI 2002).

Cambodia has twenty-four provinces (Fig. 2.1). The capital Phnom Penh is not a province but an autonomous municipality which is equivalent to a province for all practical purposes. In terms of governance and administration, Phnom Penh is treated very similarly to any other province. Each province is administered by a governor who is nominated by the Ministry of the Interior, and has to be approved by the Prime Minister.

Each province is divided into districts. The districts of Phnom Penh are called *Khan*. They are further divided into *Khum* (communes) or *Sangkat* (quarters). In Phnom Penh, *Sangkat* is used rather than *Khum*.

All administrative units, ranging from provinces and districts have their own councils. They are indirectly elected and have a mandate for five years. The councils are de facto sub-national governments. They have the authority to take legislative and executive decisions.

The councils report to the Ministry of Interior of the national government. A board of governors is established at provincial, Phnom Penh, and also at district or *Khan* levels. The chairperson of the board of governors is called Governor and the other members are called sub-governors. Governors of the provinces and Phnom Penh are

Fig. 2.1 Provinces of Cambodia

appointed by a Royal Decree, based on the request of the Prime Minister, following recommendations from the Minister of the Interior. Governors and sub-governors can give their opinions during council meetings, but they are not entitled to vote.

National Committee for Democratic Development (NCDD) is an important institution for administration and governance of Cambodia. NCDD at sub-national level is established by a Royal Decree based on the request of the Prime Minister, following nominations by the Minister of the Interior. NCDD members come from the appropriate ministries and institutions of the Royal Government. NCDD, at sub-national level, decides the responsibilities and functions of ministries, institutions, departments and authorities at all levels to identify functions to be transferred to the sub-national councils.

The capital, provincial, municipal and district councils have their own budget. This is referred to as the budgets for the sub-national administrators. *Khan* and *Sangkat* in Phnom Penh have their individual budgets specifically included in the overall budget of the Capital.

The capital councils, provincial councils, and district councils have the right to receive income from local, national and other sources of revenues in accordance with the Law of Financial Regime and Management of Assets of National Administrations. *Khan* and *Sangkat* councils have their own individual budgets to carry out their functions and duties. However, these budgets are included in the budget of the capital councils.

Local sources of revenues include, inter alia, local taxes, other non-tax revenues such as fees and charges, and voluntary donations. Revenues of the district councils are shared with the Commune and *Sangkat* councils within the district. Sub-national taxes are established by a law that is in compliance with the Fiscal Law of Cambodia. Councils also get funding from the national government. These could be conditional or unconditional funds and are available in annual disbursements.

Among the national and sub-national governments at various levels, only the national government is directly elected by the people for a maximum period of five years.

The roles and the responsibilities of the Phnom Penh Capital Council (Municipality of Phnom Penh) include legislative roles such as preparing legal documents for construction activities, issuing driving licenses and other necessary administrative functions. The roles and responsibilities of the *Sangkat* government include providing local services such as issuing death and birth certificates and preparing local development plans.

2.1.3 Cambodian Economy

Cambodia's per capita GDP in current USD was only about $78 in 1974. During the Khmer Rouge rule, the country did not produce any national statistics. Thus, no information on per capita GDP is available from 1975 to 1992. In 1993, its per capita GDP was $254. From 1998, its per capita GDP increased dramatically from $269 to

$1510 in 2018. Figure 2.2 shows changes in the nation's per capita GDP from 1960 to 2018.

Annual GDP growth rates, between 1994 and 2018 are shown in Fig. 2.3.

Over the past two decades, between 1998 and 2018, Cambodia witnessed robust economic growth rates which averaged about 8%. During this period, Cambodia was one of the fastest increasing national economies of the world.

The country achieved a lower middle-income status in 2015. The aspirations are that it may attain upper middle-income country status by 2030.

With high economic development growth rate, poverty rates have steadily declined from 47.8% in 2017 to 13.5% in 2014, a fall of 34.7% in only four years. About 90% of the poor live in rural areas.

The Gross National Savings rate of Cambodia for 2018 was at 24.4% of GDP, compared to 21.5% in 2017 (ADB 2019). Government savings were little over 10% and rest were private savings. Current account deficits increased to 10.4% of GDP in 2018, from 9.7% in 2017 (World Bank 2018). However, this deficit was fully financed by foreign direct investments (FDI) which reached a record high of 13.4% of GDP (World Bank 2019). The private sector credit grew in the range of 24.4%–31.2% during the 2011–2017 period. Public Finance of Cambodia continues to be weak, with its revenues expected to be 20.5% of the GDP in 2017, including 2.6% of the GDP as grants. The expenditure was expected to be 23.4% of GDP in 2017. The expenditure has been higher than revenue in the range of 1.3–4.1% during the 2011–2017 period. Thus, the government of Cambodia has been borrowing every year to cover its national expenditure. The exports of Cambodia have grown from

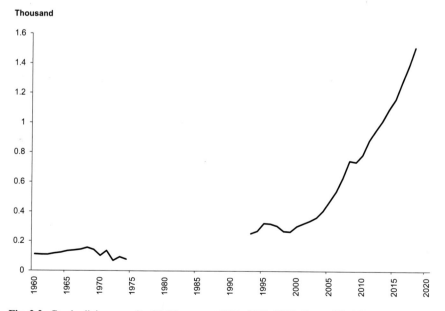

Fig. 2.2 Cambodia's per capita GDP in current USD, 1960–2018. *Source* World Bank (2019)

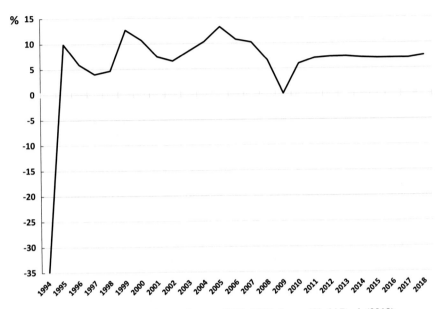

Fig. 2.3 Cambodia's annual GDP growth rates, 1994–2018. *Source* World Bank (2019)

USD 5.03 billion in 2011 to USD 10.34 billion in 2017. Imports have also increased, from USD 7.73 billion in 2011 to USD 14.51 billion in 2017. The current account deficit of Cambodia was 9.4% of GDP in 2017.

Economic growth in Cambodia has been good in recent years due to robust textile and footwear exports, real estate and construction-related activities, and strong domestic demands. Country-wise, China has made the highest foreign direct investment into Cambodia. Chinese companies have invested into the garment sector to take advantage of Cambodia's preferential tariffs with the European Union, as well as due to lower cost of manufacturing in Cambodia compared to China.

Fiscal deficit of Cambodia widened from 1.6% of GDP in 2017 to 1.9% of GDP in 2018. There is wage pressure in the system and the wage bill is expected to increase by 0.2% of GDP and non-wage current expenditure to go up by 1.0% of GDP. This has resulted in pressure on public finances due to steady increases in wage bills. The compensation to employees has gone up from 4.4% of GDP in 2011 to 7.4% in 2017 and is expected to go up to 8.1% by 2021.

The total projected revenue of Cambodia in 2016 (projected) was USD 3.69 billion,[1] with 94.4% of revenue coming from the Central government sources. The direct tax collection in 2016 was USD 752 million and total indirect tax collection was USD 2,230 million, of which USD 575 million came from trade taxes. The annual capital expenditure in the 2011–2016 period ranged between USD 1.14 billion to USD 1.61 billion. More than 75% of the capital expenditure was financed by external sources.

---

[1]Conversion rate: 1 USD = KHR 4,000.

The external public debt of Cambodia, including arrears, stood at USD 6.45 billion by the end of 2016. This was about 32% of the GDP. Corresponding figure for 2008 was 27% of the GDP. The multilateral loans accounted for 29.3% and the bilateral loans comprised 70.7% of the external loans. China is the largest bilateral creditor, accounting for around 70% of the total bilateral debt (IMF 2017) as of end of 2016. The International Monetary Fund has proposed a ceiling of contingent liabilities under Public Private Partnerships to 4% of GDP. The IMF also concluded in 2017 the country remained at a low risk of debt distress (IMF 2017).

The official currency of Cambodia is the Cambodian Riels (KHR), though the USD is widely used as a concurrent currency for national, regional and local transactions.

2.2 Phnom Penh

2.2.1 Population

In 1950, the population of Phnom Penh was 334,000. It is estimated that it was around 457,000 before the military coup in 1970. The deliberate policy of the Khmer Rouge regime was to reduce the population of all urban areas, including Phnom Penh. By 1978, Phnom Penh's population had declined to less than 50,000. This was the highest depopulation that any city has ever witnessed in the entire human history due to deliberate government policies and their strict enforcement.

After the fall of the Khmer Rouge regime, Cambodia started to recover, economically, socially and politically. By 1990, following dramatic changes in political, economic and social conditions, the population of Phnom Penh had increased manifold to over 600,000. Since then, with increasing urbanisation and industrialisation, the population of Phnom Penh has increased steadily.

Many cities in the developing countries, including Phnom Penh, have informal settlements. The number of people living in such informal settlements is not often reliably captured by the official population census estimates. Therefore, the official estimates of the population of any city, including Phnom Penh, are not authoritative. In 2013, the Ministry of Interior, estimated Phnom Penh's population to be 1.39 million. However, it is believed that the actual population of Phnom Penh may have been much higher due to the migrant labourers living in Phnom Penh. The informal estimate of the population of Phnom Penh was estimated to be around 2.0–2.5 million in 2017. Population of Phnom Penh is rather young, with 46% of the population being less than 24 years of age.

Changing population numbers of Phnom Penh over the period 1950–2020 are shown in Fig. 2.4.

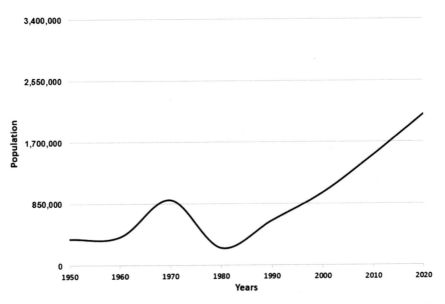

Fig. 2.4 Phnom Penh Population, 1950–2020. *Source* World Population Review (2020), UN (2019)

2.2.2   Area of Phnom Penh

The area of Phnom Penh Municipality has increased from 376.2 km^2 in 2008 to 678.46 km^2 by 2013. In 2008, the number of *Khans* (districts) in Phnom Penh increased from 7 to 8, with the division of Russey Keo District into two: Russey Keo and Sen Sok. In 2010, 20 communes from the neighbouring Kandal province were integrated into Phnom Penh. Kandal province surrounds Phnom Penh from all directions. The merger of 20 communes of Kandal province with Phnom Penh increased Phnom Penh's area to 678.46 km^2. In 2011, Dangkor district was further divided into Dangkor and Por Sen Chey. Similarly, in 2013, Meanchey district was divided into Meanchey and Chbar Ampov. Russey Keo district was further divided into Russey Keo and Chroy Changvar, parts of Sen Sok and Por Sen Chey districts were combined to create Prek Pnov District. Details of transition of Phnom Penh area and number of *Khans* and districts are shown in Table 2.1. *Khans* of Phnom Penh are shown in Fig. 2.5.

The average population density of Phnom Penh is 2100 people/km^2 or 21/ha. Population densities across *Khans* of Phnom Penh vary from a low of 5/ha in Khan Prek Pnov to 338/ha in *Khan* 7 Makara (PPWSA 2017).

Table 2.1 Expanding areas and *Khans*, Phnom Penh, 2008–2013

	Area (km^2)	*Khans* (districts)	Remarks
2008	376.2	8	• Russey Keo was divided into Russey Keo and Sen Sok
2010	678.46	8	• 20 communes were added
2011	678.46	9	• Dangkor was divided into Dangkor and Por Senchey
2013	678.46	12	• Meanchey was divided into Meanchey District and Chbar Ampov • Russey Keo was divided into Russey Keo and Chroy Changvar • Part of Sen Sok and Part of Por Senchey were divided to create Prek Pnov

Source PPWSA

2.2.3 Physical Conditions

Phnom Penh is located to the west of the X-Shaped confluence of Mekong, Tonlé Sap and Bassac Rivers. The terrain slope is from the north-west to the south-east. The elevations range from 5 to 15 m above the mean sea level. Yearly average rainfall is 1400 mm. Annual rainfall is very unevenly distributed. It rains less than 5 mm per month in January and February, and more than 200–300 mm in September and October. Yearly evaporation is about 1016 mm and ranges from a low of 56 mm in September, to a high of 123 mm in March. There are limited groundwater resources, or artesian aquifers, in the Phnom Penh region.

2.2.4 Economy

Two main pillars of the Cambodian economy at present are garment and construction industries. They are mostly located in Phnom Penh and its surrounding areas. As per economic census of 2011, Phnom Penh had 20% of Cambodia's commercial establishments and 96,000 commercial establishments, which accounted for nearly one-third of total employment in the country.

According to the National Institute of Statistics (Cambodia Social and Economic Survey 2009–2013), 80% of the population in Phnom Penh had disposable income of less than USD 129 per month. (Fig. 2.6).

Nearly one-quarter of the people working in Phnom Penh are employed in the public sector. A similar percentage of the people (around 26%) of the total workforce are employed in the private sector. As late as 2011, around 14% of the people earned their livelihoods from farming-related activities.

In recent years, like in many other Asian developed and developing countries, income disparities of the people living in urban areas are becoming wider and wider. Rich people are getting even richer, and the incomes of the poor are increasing marginally. For example, about a decade ago, it was somewhat unusual in Phnom

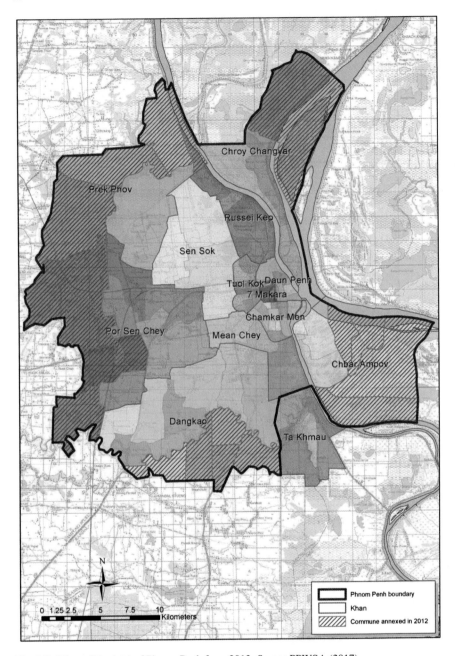

Fig. 2.5 *Khans* (Districts) of Phnom Penh from 2013. *Source* PPWSA (2017)

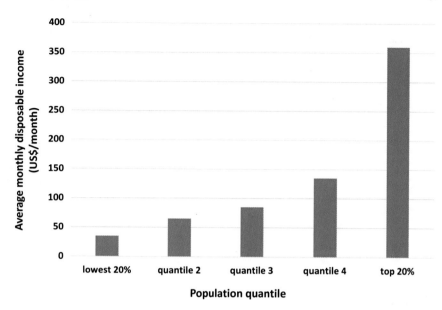

Fig. 2.6 Disposable income distribution in Phnom Penh, 2013. *Source* PPWSA (2017)

Penh to find luxury car dealerships. Now they are all over the main streets. Many of them sell around 40 expensive cars each month. Similarly, the city has seen a proliferation of construction of new luxury and expensive apartments during the past decade. With increasing urbanisation, industrialisation and robust economic growth rates in recent years, numbers of high net worth individuals have increased steadily.

References

ADB (Asian Development Bank) (2019) Cambodia, 2019–2023—inclusive pathways to a competitive economy. Asian Development Bank, Philippines

IMF (International Monetary Fund) (2017) Cambodia: staff report for the 2017 Article IV consultation—debt sustainability analysis. https://www.imf.org/external/pubs/ft/dsa/pdf/2017/dsacr17325.pdf

IWMI (International Water Management Institute) (2002) Drinking water quality in Cambodia Country report. https://publications.iwmi.org/pdf/H033491.pdf

PPWSA (Phnom Penh Water Supply Authority) (2017) Phnom Penh Water supply authority third master plan, period 2016–2030, vol 1 (Master Plan Report). Phnom Penh Water Supply Authority, Phnom Penh

UN (United Nations) (2019) World population prospects 2019. United Nations, New York

World Bank (2018) Cambodia economic update: recent economic developments and outlook. World Bank, Washington DC

World Bank (2019) Data: Cambodia. World Bank National Accounts Data, and OECD National Accounts data files. https://data.worldbank.org/indicator/NY.GDP.PCAP.CD?locations=KH

World Population Review (2020) Phnom Penh Population 2020. https://worldpopulationreview.com/world-cities/phnom-penh-population/

Chapter 3
Phnom Penh Water Supply Authority

3.1 About Phnom Penh Water Supply Authority

The Phnom Penh Water Supply Authority (PPWSA) is responsible for supplying drinking water to Phnom Penh. It was formally established in 1959 under a Royal Decree as a state-controlled business under the direct supervision of the Phnom Penh Municipality. However, the piped drinking water supply in Phnom Penh started well before 1959. In 1895, Compagnie des Eaux d'Electricité de l'Indonchine (CEEI), a French water and electricity authority in Indochina, started distributing treated water to many of the Phnom Penh residents. The same year, Phnom Penh's first water treatment plant, having a capacity of 15,000 m^3 per day, was completed at Chroy Changvar. A water distribution pipeline of 40 km was laid in Khan Daun Penh in the eastern part of Phnom Penh around 1895. Two water towers were built in Chroy Changvar in 1895. They were subsequently renovated in 1999 and are still being used.

3.1.1 1959–1970: Period of Expansion of Water Services

On 24 January 1960, King Norodom Sihanouk passed a royal decree which divided CEEI into two new institutions, one for water and the other for electricity. The new water institution responsible for domestic water supply of Phnom Penh was called Régie des eaux de Phnom Penh (RDE). Electricité du Cambodge (EDC) was created for electricity supply of the capital. Much later, in 1996, RDE became Phnom Penh Water Supply Authority (PPWSA), an autonomous water utility (PPWSA 2020).

In 1958, Chamkar Mon Water Treatment Plant (WTP) was built, having a treatment capacity of 10,000 m^3 per day. During the 1959–1970 period, new water treatment plants were built and also the older water treatment plant was rehabilitated. A year later, in 1959, Chroy Changvar WTP was renovated and its capacity was increased to 40,000 m^3 per day. In 1966, another new WTP was constructed at Phum Prek, having a capacity of 100,000 m^3 per day.

In 1957, a distribution network of 36 km was constructed. Along with the commissioning of Phum Prek WTP in 1966, the water distribution network of Phnom Penh was expanded by 233 km. A total of 165 km of distribution network was laid between 1960–1970 by the then water utility, RDE.

By the end of 1970, RDE had a 288-km piped water distribution network that was entirely made of cast iron. In 1970, RDE's total water treatment plant capacity was 155,000 m^3 per day.

3.1.2   1970–1979: Period of Conflict and Instability

General Lon Nol took control of Phnom Penh after a military coup in 1970. After this coup, there were internal armed conflicts between different warring factions of the country. All the government institutions suffered from this instability, including the water utility. There was a plan to make the water utility an autonomous institution. This had to be postponed because of continuing political instabilities that existed during this period (PPWSA 2020). By 1975, PPWSA water production was 150,000 m^3 per day, and the distribution network was virtually unchanged from 1970 (Das et al. 2010).

The Khmer Rouge took control of Phnom Penh in 1975. Under the Khmer Rouge rules, the urban population of Phnom Penh were forcibly moved from the city to the rural areas to develop and practice agricultural activities. As noted earlier, the urban population was estimated to have dwindled by nearly 90%, from 457,000 in 1970 to less than 50,000 by 1979. The potable water production was available for only a limited number of influential leaders of the city. Most of the water treatment and distribution plants were mothballed by 1979 and Phnom Penh's water production was reduced to mere 65,000 m^3 per day (Das et al. 2010).

3.1.3   1979–1993: Period of Turbulence

After the defeat of the Khmer Rouge, on 7 January 1979, by the Vietnamese armed forces, a new regime was installed in Phnom Penh. Though the Khmer Rouge had been relegated to a small area around the periphery of Cambodia, because of the then prevailing geopolitical reasons, the United Nations continued to recognise the Khmer Rouge as the official government of the Democratic Kampuchea till 1982. Thereafter, the United Nations considered the Coalition Government of Democratic Kampuchea as the official government till 1993. Again, due to geopolitical conditions and rivalries, several Western nations imposed trade embargos on Cambodia which seriously constrained its social and economic development.

The Khmer Rouge policies of forcefully moving educated and skilled manpower from urban areas to work in agricultural fields in rural areas not only ensured serious depopulation of Phnom Penh but also all its institutions, including schools, colleges, hospitals and water and electrical supply services became for all practical purposes dysfunctional.

It was a complex and very difficult task to make the water utility of Phnom Penh functional again. Years of total neglect of management, operation and maintenance of the system, lack of skilled and trained manpower, destruction of all available records of water supply systems, lack of funds to buy spare parts and necessary materials, shortage of electricity to operate the plants, and rampant corruption made any improvement to the water system extremely difficult.

By 1979, Phum Prek WTP could produce only 50,000 m^3 per day water, half of its designed capacity of 100,000 m^3 per day (PPWSA 2020).

In 1983, Chroy Changvar plant was restarted with a capacity of 40,000 m^3 per day (PPWSA 2020). However, lack of electricity and requisite resources, Chroy Changvar production came to complete halt by 1984.

In 1986, the necessary authorisation was received by the water utility to collect fees from the households for the water consumed. However, only about 40% of the customers paid their bills and the total revenue could cover only about half of its operational expenses (Das et al. 2010).

Institutionally PPWSA was in a limbo during the 1980s. From January 1988 to June 1991, it was supposed to be managed and administered as an autonomous public enterprise. However, it regularly lost significant amount of money each year. Consequently, its autonomous status had to be suspended in June 1991. It was then placed again under the administrative and budgetary control of the Phnom Penh Municipal Authority (Biswas and Tortajada 2009).

In 1992, PPWSA had an annual deficit of KHR 809 million, which meant that the two-third of its expenditure had to be subsidised by the Phnom Penh Municipality. PPWSA had a poor water bill collection rate of only 19%. In 1992, the domestic tariff was raised to KHR 166 per m^3. The changes in domestic water tariffs in Phnom Penh from 1981 to 1993 are shown in Fig. 3.1.

Even after this tariff increase in 1992, PPWSA operating ratio in 1993 was 150%, implying that its costs were 150% of its revenue (Colon and Guérin-Schneider 2015). At that time, major and important public sector customers like the army, various government departments, and influential people like the ministers never paid their water bills (Biswas and Tortajada 2009).

Numerous households that received piped water supply used to resell water. The cost of water re-sold by these households and the private vendors was several times the cost of water supplied by the PPWSA (Biswas and Tortajada 2009).

PPWSA then lacked technical, managerial and administrative capacities and proper equipment to rehabilitate the ageing and ailing water infrastructure (Dany et al. 2000). There were no records of who its consumers were and how much water they were using. Nor were maps available as to where even the pipes were laid. All available records were destroyed during the Khmer Rouge rule.

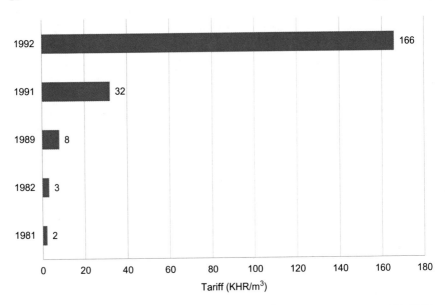

Fig. 3.1 Domestic water tariffs in Phnom Penh (KHR per m^3), 1981–1992. *Source* PPWSA (2009)

The institution was managed in a completely bureaucratic and an inefficient manner. Nearly 80% of the workforce of PPWSA worked for less than two hours a day, and that too inefficiently and without any proper guidance and supervision from its managers. The workforce had limited skills and no clearly defined job descriptions. Even though it was an autonomous institution, it was for all practical purposes run as a government department, with no administrative, operational and financial autonomy or any type of accountability. It needed continuous and regular municipal authorisation for all its operational expenditures.

Getting any authorisation was a time-consuming process during which the underlying problems deteriorated even further. Whatever revenues the water utility earned were consolidated within the general municipal funds. Thus, there was no incentive on the part of the PPWSA management to run the institution as a standalone efficient business. Neither were any legal or institutional frameworks available for it to carry out its functions in a reasonably efficient manner (Biswas and Tortajada 2009).

By the early 1990s, PPWSA as an institution, as well as its overall management was dysfunctional and in total shambles. Consequently, and not surprisingly, consumers received very poor and unacceptable levels of service. For example, in 1992, it had only three technicians who were capable of connecting meters. Salaries of its staff were low which did not allow them to maintain their families. Thus, the institution was riddled with corruption from top to the bottom.

By the end of 1993, PPWSA had a total of 26,881 piped water connections, with a collection ratio of only 48% and non-revenue water at an estimated 72%. The length of the potable water distribution network was 288 km (Das et al. 2010). Theoretically only about 20% of the Phnom Penh residents had access to piped water, and that too for a limited period each day (Otis 2013). The piped water infrastructure was in poor shape and needed urgent rehabilitation for which neither adequate funds nor technical expertise were available (Curtis 1993).

3.1.4 1993–2012: Transformation Phase

The transformation of PPWSA from beleaguered operations in 1993 to a path of healthy management started with the changing of the guards when Ek Sonn Chan was appointed as its Director General.

Coincidentally, 1993 was also a landmark year in the history of Cambodia since its Government was formally recognised by the United Nations. This formal recognition meant that trading embargoes imposed by several Western nations were lifted. This lifting quickly led to involvements by certain major donor agencies of the Western world in the country's development process. For the first time a concerted attempt was made by a few donor agencies to improve the social and economic conditions of the country, including improvements in Phnom Penh's water supply services.

The combination of the charismatic and efficient management leadership of Ek Sonn Chan and the donor support helped PPWSA transform itself from a loss-making, corrupt and inefficient water institution in 1993 to become one of the most efficient utility operations, with a healthy bottom line, for any utility in a developing country. What is also noteworthy is that this remarkable transformation occurred in only about a decade.

Ek Sonn Chan: a charismatic and dynamic leader

Ek Sonn Chan came from a poor and humble Cambodian family. He had to work hard to support his own studies. He received his primary education from the POEU OUM primary school in Phnom Penh. His secondary education occurred in two places, at College of Kampong Thom, in the district of Kampong Thom, located 127 km west of Phnom Penh, and at the College Intradevy in the district of Tuol Kork, Phnom Penh.

In 1974, he graduated with a degree in electrical engineering from the Institute Technique de l'Amitié Khméro-Soviétique (Superior Technical Institute of Khmer-Soviet Friendship). Currently, this institution is known as the Institute of Technology of Cambodia, one of the country's highest and the best educational institutions in Phnom Penh. His first job, after completing his electrical engineering studies, was with a French rubber plantation factory near Phnom Penh. He had to move out of Phnom Penh during the Khmer Rouge regime. After Khmer Rouge was ousted, he returned to Phnom Penh and received his first government job as a butcher in a government owned abattoir, even though he had graduated in electrical engineering.

A serious, unassuming and hardworking man, he showed leadership qualities from the beginning of his career. He was steadily promoted and became the Deputy Chief of Commerce of the Phnom Penh municipality. In 1983, he became the Director

Fig. 3.2 Ek Sonn Chan.
Source PPWSA

General of Commerce for the municipality. In 1989, he became the Director of Electricity Authority of Cambodia (EAC), the utility that was responsible for providing electricity to the country. On 11 September 1993, Ek Sonn Chan was appointed as the Director General of PPWSA. This is when the real Phnom Penh water story begins (Fig. 3.2).

When he was appointed as the Director General of PPWSA, he had limited knowledge of running a water utility, let alone much experience on how to transform a non-functional utility like PPWSA to an efficient and financially sustainable institution. Following his appointment as the Director General, he spent considerable time learning how an efficient water utility should be run and successfully managed.

Fortunately for PPWSA, Ek Sonn Chan turned out to be a fast learner and a very competent leader. He provided good and stable leadership to the utility for nearly two decades, from 1993 to 2012. He recruited a dedicated, committed and honest senior management team. He successfully created and nurtured a culture of performance as well as cooperation among all the employees of PPWSA. The utility, for the very first time in its history, started to function as a team committed to achieve its goal, that is to provide continuously and reliable clean water to all the households of Phnom Penh on a 24 × 7 basis.

As the incoming Director General, Ek Sonn Chan faced numerous complex and difficult challenges. His major challenges included improving the financial viability of the utility, reducing corruption that was rife within the organisation, dismantling unauthorised piped connections which were widely prevalent all over the city, replacing old pipes with new ones, and rehabilitating and updating various water production facilities (Chan 2009).

Unauthorised connections to the utility's network was a pernicious and widespread practice. Unfortunately, this practice is still widely prevalent in numerous cities of the developing world. These connections were often made by the corrupt PPWSA technicians, during working hours using utility's instruments, for personal financial gains. An example of illegal connection that existed in Phnom Penh is shown Fig. 3.3.

Fig. 3.3 Unauthorised piped connections. *Source* PPWSA

In 1993, PPWSA had minimal technical, administrative, financial and managerial capacities to provide basic water services. It had no funds even to carry out the basic functions that are considered to be absolutely essential for any water utility to provide a minimally acceptable level of service. For example, it had no alum to treat water even to a minimum standard. Whatever alum used to be available, were mostly stolen by its staff, and then sold privately in the market. The institution also had no money to pay for its electricity bills. Fortunately, this was not really an important issue since electricity was available free at that time because PPWSA was a state-owned entity. However, alum had to be bought from the private sector but the utility had no financial resources to pay for it. The Director General approached several donors for funds to buy the necessary but essential alum. Finally, the Japanese Government agreed to provide $50,000, for only the year of 1993 in order that the utility could buy alum and chlorine from the open market. This allowed the utility to provide reasonable quality of water for consumption by its customers. Ek Sonn Chan undertook many other steps that led to transformation of PPWSA from a financially and operationally struggling water utility to a successful and healthy water utility. Some of the key steps and reasons for the institution's transformation under his leadership are discussed next.

Consumer Data Base—Shortly after becoming the Director General, Ek Sonn Chan realised there was no information available on who were its customers. It also had a dismal record of collecting payments from its customers who were billed for water they used. In addition, most of the staff members of PPWSA never paid their water bills. In fact, many of them, including even the earlier Director General, sold water they received free from PPWSA to their neighbours. Equally, some of the major public sector users of water, like the army, never had any water meter installed. They, thus, never paid for the water they consumed, even though it was the utility's largest water consumer. Equally, various government departments, ministers and many influential people seldom paid for water used by them, whether billed or not.

A survey to determine PPWSA's customer base was initiated in 1994 by the newly appointed Director General. This survey was initially expected to be carried out by a French consulting company, with financial aid from the French Government. However, when the Director General realised that this 2-year contract with the French consulting company would provide customer information for 50% of *only* one district (*Khan*), he requested the Mayor of Phnom Penh municipality to give him 100 people to complete the entire task in one year. Since there were many people who were underemployed in the municipality, the Mayor promptly acceded to this request. The newly recruited staff were then trained to conduct the survey properly and reliably. An important initial success of the utility was that its management not only trained the people well but also completed the survey within the stipulated one year. The French consulting company soon realised that with only 10 people they could not compete with the PPWSA team in terms of completing the work they were expected to do. They then stopped their work (Biswas and Tortajada 2009).

The findings of the survey were revealing. It found that there were 13,901 households in Phnom Penh who were getting water but were receiving no bills. Equally there were many households who were receiving bills but no water!

Following this successful survey, the French Government gave a grant to PPWSA to establish a fully computerised and up-to-date database of its customers, and then update the database regularly. The database became fully operational in 1996.

The computerised data system was further expanded in 2001 to handle all financial transactions and operations of PPWSA. By 2003, a reliable and comprehensive financial management information system was in place. This enabled the institution to have immediate access to the latest and historical financial data and its revenue collection status in real time. An indirect benefit of this automated information system was that the corruption and abuse of power for bill preparation and collection were significantly reduced, and for the most part eliminated. Following this success, the Authority has continued to improve its overall financial management practices on a regular basis over time.

Unaccounted for water: An employee-centric approach—In 1993, a major problem PPWSA faced was very substantial water losses due to high unaccounted for water. This was then estimated to be over 70% of total water production. Reducing unaccounted for water losses within a reasonably short timeframe required that several tasks had to be undertaken and completed almost concurrently. This required a strict system approach. No water supply can be kept at a reasonable and affordable level when the incomes from the customers do not cover the costs of providing a good and reliable service. In the case of PPWSA, nearly three-quarters of the water produced in 1993 yielded no revenue whatsoever. Thus, a strict regime to reduce these high losses was planned and implemented. This had several interrelated components.

The new Director General realised that no significant reduction of unaccounted for water was possible without the cooperation and the support of dedicated, competent, motivated and honest staff members. During the early 1990s, the quality of the staff members in PPWSA left much to be desired. Not only the staff members were demoralised but also they had very good reasons to be demoralised because of poor governance practices, lack of clear job descriptions, below subsistence level salaries, lack of discipline, absence of incentives for good performances, and pervasive corruption over the years which had sapped morale of the workers. Lethargy, poor working conditions and practices, uncaring and not properly trained staff and absence of any accountability, ensured consumers received a poor service. Accordingly, the overall work culture of the utility had to be changed by enforcing strict disciplines in a sensitive, fair and transparent manner. This was a difficult task since the rest of the public sector employees in Cambodia were facing very similar situations, and also were behaving in a very similar manner.

It was a difficult and complex task to change the then existing institutional culture of PPWSA. It had to begin with the senior officers who had to become role models for the rest of the employees. Earlier, one of the perks of the job was that the employees of PPWSA received free supply of water. This practice was immediately stopped. Staff members not only had to install meters in their houses but also had to pay their water bills in full, like any other normal inhabitant of Phnom Penh, and within the stipulated time period. Otherwise, they were treated in the same way as those who were delinquent with their bills and had to pay penalties to ensure continued access to water supply.

Ensuring discipline and honest behaviour from the PPWSA staff took some time. In retrospect, a strict enforcement of the rules and a "carrot and stick" approach proved remarkably successful. Initially, as to be expected, there were some stiff resistances. For example, some senior staff continued with their corrupt behaviours, or had no interest in obeying the new rules and discipline. One senior staff member felt that since he came from a powerful family and was well-connected politically, he was untouchable and could do whatever he wanted and whenever he wanted. When he was dismissed for his consistent poor performance and corrupt behaviour, he threatened to sue and kick up a political fuss. With overwhelming evidence against him, he finally had to back down and was promptly dismissed from PPWSA. The fact that a politically well-connected person was dismissed for corrupt practices promptly, in a fair and transparent manner, sent an important signal to the rest of the PPWSA staff members. They all realised that there were new working rules where if they abused their position and power, especially in terms of corrupt practices, like providing illegal connections, incorrect meter readings, or accepting financial contributions for any unofficial purpose, they would be promptly dismissed after transparent disciplinary hearings.

Simultaneously other rules and incentives were put into place. These were strictly and fairly enforced in a prompt, clear and transparent manner. One good example is that if a meter reader of an area did not, or could not, find an illegal connection, but one of his colleagues did, the colleague concerned received a reward, and the meter reader was penalised. The general public was made aware of the problems of illegal connections and how they endangered the quality of service that may receive. Those customers found to have illegal connections were heavily penalised, and anyone who reported an illegal connection was rewarded. Inspection teams were set up to search for, find and then promptly eliminate all illegal connections as soon as they were found.

As a result of these and other associated actions, the number of illegal connections discovered dropped from an average of one per day, when the operation started, to less than five per year by 2002. The process was further helped by incentives to employees who contributed to the reductions in the levels for unaccounted for water. As PPWSA became increasingly viable financially, it could increase the salaries of its employees by around ten times over the 1994–2004 period. Furthermore, employees were trained and rewarded for their good work, and equally penalised for poor performance. All these efforts proved to be immensely successful in reducing unaccountable water from over 70% in 1993 to less than 10% by 2012.

Elimination of underground tanks—PPWSA suffered from the prevalence of underground tanks. Most of the underground tanks had been constructed with water sourced from illegal tapping from water distribution pipelines, as shown in Fig. 3.4.

After the fall of the Khmer Rouge regime, water supply was erratic and the pressure was too low to supply water to higher demands from a steadily increasing number of households. People who returned to Phnom Penh after the Khmer Rouge atrocities had no option but to construct underground tanks because that was the only way they could obtain their daily water needs. It was estimated that there were around 1945 underground tanks in Phnom Penh in 1993.

Elimination of the underground tanks received priority attention from the new management. PPWSA decided to determine the magnitude of the problem posed by these underground tanks by conducting a survey in 1994. Shortly thereafter, a policy

Fig. 3.4 An unauthorised underground tank, where a pump was used to draw water. *Source* PPWSA

decision was taken to close all such tanks as quickly as possible. All the house-holds that depended on such tanks for access to water received individual house connections, but with meters. When this was not possible for a few specific cases, a caretaker was nominated for each tank. All connections from the tank were provided with meters. The caretaker was often a local vendor-cum-retailer and had to pay to the utility for the total water used from the tank at the prevailing domestic rate. The caretaker then collected the funds from the consumers. PPWSA ensured the care-takers received reasonable net incomes for their work, after paying the appropriate water charges. By taking various similar measures, all the underground tanks were eliminated by 2003.

Metering—In order to ensure that a transparent and a fair system exists, it is essen-tial that all the connections should be metered to determine reliably actual water consumptions of individual households. Only after each connection is metered, households could receive accurate bills which directly reflected the amount of water they consumed during a specific period.

In 1993, out of a total of 26,881 house connections, only 3391 connections were metered. In other words, more than 87% of the households connected to the network were receiving estimated bills. These often had no linkages to the quantities of water that were consumed by individual households. After taking a policy decision to move to a system where all connections would be metered, and as soon as possible, the number of meters installed increased steadily. As a corollary, the number of unmetered connections started to decline. By 2001, all the household connections were metered. In addition, over time, more accurate Class C meters were installed to replace less reliable Class B meters. These developments further increased the income and credibility of PPWSA over time.

Upgrading Infrastructure—In 1996, the PPWSA started to rehabilitate its water network with funding support from the Asian Development Bank, the World Bank and the Governments of France and Japan. By 1999, all the old cast iron pipes were replaced. The rehabilitation process was completed by 2002. It included construction of a new 16-km water transmission line with a loan from the Asian Development Bank. In addition, a maintenance and repair team was organised on a 24-h standby basis. The public was encouraged to report all leaks which were then promptly repaired.

Autonomy of PPWSA—Before 1996, water supply of Cambodia was regulated by the Ministry of Public Works and Transportation. This Department had offices in each province and in Phnom Penh which managed water supplies in respective areas. Phnom Penh Water Supply reported to this Department in Phnom Penh, which, in turn, reported to its parent ministry and the Phnom Penh Municipality. Investment plans for Phnom Penh had to have at least three steps before they could be approved (Fig. 3.5). Three different institutions had to approve any investment plan: Depart-ment of Public Works and Transportation, Phnom Penh Municipality and Ministry of Economy and Finance. The approval process was bureaucratic, cumbersome, rigid,

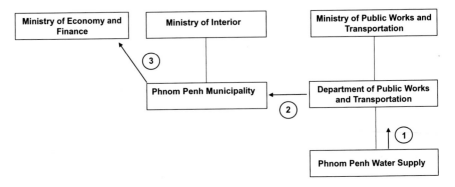

Fig. 3.5 Decision making of PPWSA's investment and other plans prior to 1996

and time-consuming. It was thus inefficient in terms of planning, operation, and overall management. The results left much to be desired.

In a radical move, PPWSA was granted full autonomy under sub-decree Number 52 of 19 December 1996. An unofficial translation of this sub-decree is provided in Annex I. As per this decree, the new autonomous entity responsible for water supply in Phnom Penh was renamed Phnom Penh Water Supply Authority (PPWSA). The Authority was no longer required to report to the Department of Public Works and Transportation (Phnom Penh office) and it came under direct control of the Phnom Penh Municipality. In a significant departure from the past, a sub-decree mandated the administrative management of the new institution to a newly established Board of Directors. According to this sub-decree, the Director General is responsible for its day-to-day operations. This person is appointed for a 3-year period, but can be reappointed to this position for any number of additional terms of three years. The Director General is appointed by the Prime Minister, after being nominated from its tutelary ministry.

The decree stipulated that PPWSA must organise, manage, and operate all its activities independently in accordance with commercial business requirements (Article 4). It could have an independent financial compensation package for its staff, as long as it is approved by its Board of Directors (Article 10). The Director General will have the authority to hire and fire staff (Article 14). Furthermore, the Director General must submit to the Board of Directors an annual plan each year before 1st October. This annual plan is required to cover the investment and financial plans, operational budget, price of water and other services to ensure that the total revenue is adequate to cover the utility's operational expenses. The autonomy mandated that the annual plan of PPWSA must be approved by the controlling ministries within three months of its submission.

PPWSA came under direct control of the Ministry of Industry, Mines and Energy of the Royal Government of Cambodia from 2004 onwards. It was no longer required to report to the Phnom Penh Municipality. From 2014, PPWSA has been reporting to the Ministry of Industry and Handicrafts which is now called the Ministry of Industry Science Technology and Innovation.

Overall, autonomy granted to PPWSA from 1996 onwards was one of the most important drivers for the steady improvement in the operational performance efficiencies of PPWSA.

Improvement in quality of financial reporting of PPWSA—In 1990, Cambodia had a socialist regime and the operations of PPWSA were primarily dependent on financial subsidies received from the Government. The financial accounting under the socialist structure did not require preparation of any annual profit and loss statements, or balance sheet assessments. The accounts were prepared under just twelve main accounting headings. Essentially the financial accounts of that time captured only two main features, first, from where the funds were received, and second, how they were spent for different items.

Before 1993, Cambodia had no financial accounting standards. After the first general elections in Cambodia, in 1993, the Ministry of Economy and Finance adopted the French system of accounting, which was very strict and comprehensive. However, in 1994, PPWSA received major aids from the World Bank and the UNDP. The aids from both these institutions were conditional. They explicitly stipulated that PPWSA had to adopt Anglo-Saxon accounting standards from thereon to receive funding support.

A UNDP sponsored Canadian expert helped PPWSA to build expertise in Anglo-Saxon accounting rules and procedures. PPWSA also had to learn how to make financial projections through financial modelling. Before PPWSA became an autonomous institution, all its financial statements had to be approved by the Ministry of Economy and Finance. However, the Ministry of Economy and Finance itself had no knowledge of Anglo-Saxon accounting rules and methods. Thus, they had to rely upon the expertise of PPWSA to understand the validity and reliability of its financial statements!

Progressively Cambodian accounting standards were introduced from 2003. PPWSA adopted them shortly thereafter. In 2012, it adopted International Financial Reporting Standards (IFRS). The accounts of PPWSA were audited by an international accounting company, PricewaterhouseCoopers (PwC), each year, from 1997 to 2012.

The strength of an organisation depends not only on improvements in its operational performances but also on the sophistication, reliability and depth of its financial reporting. The ability of PPWSA to conduct its financial reporting and financial analysis improved significantly from the early 1990s to 2012. This is a truly remarkable achievement in terms of capacity building. Very few water utilities of the cities of the developing world now have the financial knowhow and expertise of PPWSA. This knowledge has greatly enhanced its steady progress towards financial sustainability.

3.1.5 Donor's Role in Revival of PPWSA During 1993–2012

Asian Development Bank (ADB), French Development Agency (AFD), Japan International Cooperation Agency (JICA), World Bank (WB) and other developmental aid agencies have played important and significant roles in turning around PPWSA since 1993.

In 1993, JICA provided assistance to PPWSA for its first master plan (1993–2010). This first master plan document acted as the main planning document for PPWSA and it became the "blueprint" for PPWSA for its development during the subsequent years (Das et al. 2010). It also became the main instrument for donor consideration, which has left much to be desired in other countries and even in other Cambodian sectors.

In 1993, the water treatment plants of PPWSA were in a dilapidated state. JICA and the French Government provided much needed USD 5.2 million assistance for the rehabilitation of the Phnom Penh water system (Das et al. 2010). In 1994–1995, JICA supported urgent rehabilitation of transmission lines, and also expansion of the capacity of Phum Prek water treatment plant from 56,000 m^3/day to 100,000 m^3/day. In 1997–1999, JICA further stepped in to help in rehabilitating 88 km of water transmission and district network in the central areas of Phnom Penh. JICA, along with Japanese water operators, further trained PPWSA employees for regular and proper operations and maintenance practices. In 2001–2003, JICA assisted PPWSA in expanding the capacity of Phum Prek water treatment plant.

JICA further provided assistance for the formulation of the second master plan (2006–2020) in 2006. The second master plan laid the basis for subsequent expansion of water treatment plant capacity at Chroy Changvar and later on setting up of new water treatment plants at Niroth. In retrospect, JICA's role has been very important in planning, physical asset creation and capacity building of PPWSA from 1993 to at present.

The fact that JICA supported the formulation of the First and Second Master Plans and that PPWSA management was intimately and continuously associated with their formulation processes, ensured these two Master Plans guided all major developments during the 1993–2020 period. In retrospect, one of the direct but important benefits of the two Master Plans was that they effectively, albeit indirectly, coordinated the activities of all donor agencies during this entire period. Whatever assistance the donors were willing to provide had to be within the framework of these two Master Plans.

In addition, the fact that for nearly three decades JICA has continually supported overall planning, institutional strengthening, capacity building and some infrastructure development went a long way to ensure PPWSA's remarkable transformation, the type of which has so far not been witnessed in any domestic water supply sector of any city of a developing country. JICA is also currently helping PPWSA to formulate the up to date of the Third Master Plan (2016–2030). While other bilateral and multilateral development aid agencies have also provided significant financial assistance to PPWSA, it is probably JICA's long-term commitment which has been an important component that has contributed to PPWSA's success. The assistance to PPWSA has also probably been one of the most successful aid activities of JICA which has had significant long-term societal and economic impacts on any other developing country.

The World Bank has also provided capacity building assistance to PPWSA. During the 1990s, the World Bank had considerable influence on the Royal Government of Cambodia. It also provided advisory services on institutional arrangements. These contributed to increasingly greater autonomy for the utility. The law that brought autonomy to PPWSA was also responsible for the creation of the Coordinating Committee for Water and Sanitation and restructuring of the water tariffs. The World Bank also impressed upon the Royal Government of Cambodia that PPWSA operating ratio should not be more than 50%, that is, the costs incurred by PPWSA should be limited to 50% of its revenues.

In 1997, the World Bank provided funding for urban water supply to Phnom Penh and Sihanoukville. From 1998–2004, it further provided financial assistance for the rehabilitation and extension of the Chroy Changvar project.

The Chroy Changvar plant was originally built in 1895 with a French architectural design. The World Bank insisted on keeping the heritage structural design intact in the new Chroy Changvar plant. The original architectural glory of Chroy Changvar plant can be seen in Fig. 3.6.

Fig. 3.6 Chroy Changvar Plant. *Source* Pawan K. Sachdeva

The Asian Development Bank (ADB) provided loans to PPWSA between September 1997 and June 2003. It also assisted in the procurement of supply and installation of transmission mains to convey water from Phum Prek water treatment plant and new Chroy Changvar water treatment plant and the procurement of distribution mains. Based on the JICA Master Plan, different donor agencies, and also PPWSA, focused on different districts of Phnom Penh but always within the context of the Master Plans. ADB helped to improve the capacity building of staff members through what was called a "twinning program." Under this "twinning program," staff members were sent to the Brisbane water utility to work for a few months. Upon their return to Phnom Penh, PPWSA took advantage of the new knowledge obtained and used it to improve its own performance. In addition, representatives from the Brisbane water utility visited Phnom Penh to assess the progress of implementation of the new ideas of the returning staff members. They also provided further inputs to improve operational and management practices and processes of PPWSA. The "twinning program" helped PPWSA to improve its capacity considerably. The programme was so successful that in later years, PPWSA acted as the "guiding brother" to a Vietnamese water utility, very similar to how the Brisbane water utility had helped PPWSA.

ADB was also instrumental in bringing adoption of new tariff plans by PPWSA. In 1996, ADB proposed the initial tariff plan which was subsequently adopted by the water utility. This plan proposed three water tariff increases within a period of seven years, along with continuous improvements in service delivery. This process was designed to ensure there was no sudden huge jump in tariffs which its customers may find economically difficult and thus may become reluctant to pay. This could then have major political consequences.

With the enthusiastic endorsement of the higher tariff by the Asian Development Bank, which had conducted a socio-economic survey of Phnom Penh in terms of the consumer's willingness and capability to pay higher tariffs, the first increase in the tariff was introduced on 1 June 1997. The tariff increase had equally strong support from all the donors like the World Bank, as well as strong political backing at home, from the Prime Minister, Finance Minister and the Governor of Phnom Penh. This increase along with other steps concurrently taken by PPWSA to improve the efficiency of its operation practices, helped the institution to double its income.

The increase in water tariff was very carefully planned and strategically implemented. To start with a socio-economic survey of the water supply situation was carried out for the city. This survey included collecting information like how much consumers were paying for water purchased from the private vendors, and what were likely to be their reactions if these vendors were replaced by regular water supply from the PPWSA. This survey showed the willingness and capability of the consumers to pay a higher tariff than the one they were being charged, provided the consumers received significantly improved services. The results of this survey are shown in Table 3.1.

Increasing the water tariff was also a difficult and sensitive process in political and institutional terms. The PPWSA had to initiate a request for tariff revision through

Table 3.1 Socio-economic survey of water supply in Phnom Penh

Items	Category I	Category II	Category III
	Households connected with piped water supply	Water sold by neighbour	Water sold by vendors, and collected directly from rivers and wells
Percentage of households	42%	16%	42%
Average income (riels/month)	1,375,000	325,000	1,020,000
Average consumption (m³/month)	37	14	20
Average cost/m³ (riels)	280	2652	2124
Total water cost riels/month	10,360	37,128	42,480
Water cost as % of income	0.8%	11.4%	4.2%
Proposed average tariff (riels/m³)	575	400	450
Total monthly water expenditure (riels/month)	21,275	5600	9000
Water cost as % of income	1.5%	1.7%	0.9%

Source ADB 1996

its tutelary ministry. However, in order for the price increase to be implemented, it ultimately required the approval of the First and Second Prime Ministers.

The tariff was calculated after considering the total expenses of the PPWSA, including operation and maintenance costs and the depreciation of all its assets. The utility was expected to recover all its operating costs with tariffs, as well as the depreciation of all its assets, except land which generally increases in value over time. The value of its assets was to be revised every five years. ADB assisted in the introduction of block tariffs after 1997. History of water tariffs till 2001 in Phnom Penh is shown in Table 3.2.

The new tariff structure adopted in 1997 enabled PPWSA to become financially sustainable. After adoption of the new tariff structure in 1997, financial assistance from the donors to PPWSA came in the form of soft loans rather than grants. Before 1997, PPWSA received a total of USD 110 million in assistance in form of grants from JICA, French Development Agency (AFD) and World Bank (Das et al. 2010).

The French Development Agency also played an important role in providing financial assistance to PPWSA. AFD gave its first grant in 2004 and the first direct loan to PPWSA at a rate which was 160 basis points lower than the benchmark LIBOR in 2007. Subsequent loans from AFD also came at lower cost than LIBOR

Table 3.2 Changing tariff structure in riels m^3, 1983–2001

Time Period	Tariff Structure		Nature of Customer
Until 1983	Free		All
1984	KHR 166/m^3		Domestic
	KHR 166/m^3		Commercial
1993 to June 1994	KHR 166/m^3		Domestic
	KHR 515/m^3		Commercial
June 1994 to May 1997	KHR 250/m^3		Domestic
	KHR 700/m^3		Commercial
June 1997	Introduction of Block Tariffs All connections to be metered		
	Volume (m^3)	Tariff (KHR/m^3)	Nature of Customer
	0–15	300	Domestic
	16–30	620	Domestic
	31–100	940	Domestic
	>100	1260	Domestic
	0–100	940	Commercial
	101–200	1260	Commercial
	201–500	1580	Commercial
	>500	1900	Commercial
	All	940	Government Institutions
2001	Introduction of Block Tariffs All connections to be metered		
	Volume (m^3)	Tariff (KHR/m^3)	Nature of Customer
	0–7	550	Domestic
	8–15	770	Domestic
	16–50	1010	Domestic
	>50	1270	Domestic
	0–100	950	Commercial
	101–200	1150	Commercial
	201–500	1350	Commercial
	>500	1450	Commercial
	All	1030	Government Institutions

Source PPWSA

by +90 basis points. AFD was the first donor agency that gave three loans directly to PPWSA without the involvement of any other Cambodian government department.

Several bilateral and multilateral donors have played important roles in restoring physical assets and renovation of infrastructure, capacity building and public policy inputs that helped to radically transform PPWSA's performance.

The history of the loans and grants from the donors to PPWSA is shown in Annex II.

3.1.6   Listing of PPWSA Stock in 2012: A Vote of Confidence

PPWSA was the first stock that was listed on the new Cambodia Stock Exchange on 18 April 2012. 15% of the ordinary shares of PPWSA were issued at an offer price of KHR 6300 per share. The stock offering raised close to USD 20 million. The issue was oversubscribed 17-times. Of the 15% shares offered, 1.24% were offered to employees under Employees Stock Option Plan. The Ministry of Economy and Finance continued to hold the balance of 85% of the ordinary shares of the company.

Successful listing of PPWSA was a strong testimony to its sound financial reporting and accounting system. It also has a strong Securities Exchange and Investor Relations Office. This is a good example of sound internal management reporting practices of any listed company in any developing country.

As on 3 July 2020, the stock price was KHR 5660 per share implying a market capitalisation of KHR 492,250 billion or USD 123 million.[1]

3.1.7   Transformation of PPWSA Under Ek Sonn Chan

Ek Sonn Chan provided strong, efficient and long leadership. He received excellent and loyal support from a competent group of senior management team. These activities were further complemented by regular support from several bilateral and multilateral donors. All these efforts worked synergistically and were instrumental to dramatically transform PPWSA performance within a short period of little over a decade, starting from 1993. The results of this remarkable transformation can be seen from Table 3.3.

Ek Sonn Chan has been recognised for his stellar role in the transformation of PPWSA through many international awards, including Ramon Magsaysay Award in 2006 and an award from the President of France in 2010. As noted earlier, PPWSA also received the prestigious Stockholm Water Industry Award in 2010 which very specifically noted his leadership.

---

[1]Conversion rate: 1 USD = KHR 4000.

Table 3.3 Transformation of
PPWSA, 1993–2004

Indicator	Unit	1993	2004
Production capacity	m³ per day	65,000	235,000
Population coverage	%	50	85
Distribution network	Km	280	1077
Supply pressure	Bar	0.2	2.0
Daily supply duration	Hours per day	10	24
Number of connections	#	26,881	121,522
Staff per 1000 connections		22	4
Illegal connections detected	# per year	>300	<5
Metering ratio	%	12%	100%
Bill collection ratio	%	50%	99.9%
NRW	%	72%	14.1%
Profit after tax	Billion KHR	N/A	9.2

Source PPWSA (2017)

3.2 Changing of the Guard: 2012–2019

On 1 July 2012, Sim Sitha was appointed as the new Director General of PPWSA. Ek Sonn Chan became the Under-Secretary of State in the Ministry of Industry, Mines and Energy. Prior to joining PPWSA, Sim Sitha was the Director General of Sihanoukville Water Supply Authority (SWSA) during the 2003–2012 period.

He graduated as a civil engineer from the L'institut Technique Supérieur de l'Amitié Khméro-Soviétique that was established in 1964 with the help of the former Soviet Union. After the collapse of the Soviet Union, it was renamed Institute of Technology of Cambodia. For decades it has been Cambodia's premier technological institution for the training of engineers and technicians. Sim Sitha joined the Department of Public Works in Phnom Penh, where he worked from 1989 to 1993. He was then the site manager of a PPWSA treatment plant. From 1994 to 1996, he worked as an engineer for the Urban Water Supply Rehabilitation Project in Sihanoukville.

In February 1996, he became the Deputy Director of the Sihanoukville Water Supply Authority (SWSA), and then was appointed as the Director General of SWSA in June 2003.

Sim Sitha, by virtue of being Director General of SWSA, had good knowledge of all aspects of running a water utility before joining PPWSA. He had moved up the ranks in SWSA starting from a bill collector to the highest position as the Director General. He had exposure to construction, billing, planning and loan negotiations at SWSA. In 1997, the World Bank had given a loan of around USD 30 million to Cambodia. This was split between PPWSA which received 90% of the total loan, and SWSA received the balance of 10%. PPWSA and SWSA were both beneficiaries of technical assistance and capacity building programmes of the World Bank. Sim

Sitha was part of several loan negotiations with donor agencies during his stay with SWSA.

Sim Sitha had good knowledge of running a water utility. However, Sihanoukville is a small town and has only about 10% of the population of Phnom Penh. Accordingly, managing its water supply was less complex, arduous and sophisticated. After joining PPWSA, Sim Sitha took time to adjust to the requirements of running a much larger and difficult institution. Fortunately, he inherited a well-performing institution and had the added advantage of having well-trained and experienced senior management officers in place. The challenge for him during the initial years was to establish his credibility amongst his top management team and other employees of PPWSA. He was following the footsteps of a very remarkable person. He was also an outsider so far as PPWSA staff members were concerned.

On the positive side, Sim Sitha promptly realised that the qualities and capabilities of the people working in PPWSA were significantly higher than at SWSA. Though the scale of operations of PPWSA were much larger than SWSA, it was counterbalanced by having a management team of much higher competence at PPWSA compared to at SWSA.

Sim Sitha during his first term of office, 2012 to 2015, worked on knowing the culture and modus operandi of the organisation. He tried to understand the system rather than making an attempt to change it. He adopted a management style where he did not force his decisions on his senior management team. Rather, he gave them some alternative options and then allowed them how best to implement the management decisions taken. He would force his ideas on the management team members only as the last resort. He made some changes in the organisation structure of the utility and made some alterations in the individual portfolios of the senior management staff members.

After completing his first term in 2015, he made some additional changes during his second term.

His emphasis has been to improve further the technology adoption rate of the institution and human resources development. An example is though the accounting system of PPWSA was automated in 2012 the year he joined the utility, there were still some constraints in cash flow statement preparation as well inadequate interdepartmental coordination in terms of data sharing. He has worked to improve the data sharing across departments and considered improving the quality of internal reporting within PPWSA. He also made external reporting more robust, comprehensive and timely. He has further improved and updated all the standard operating procedures of all the departments.

He has introduced the adoption of a geographical information system and spot billing system. He considers the technology adoption process within PPWSA as an "internal information technology revolution." His focus has been to improve information technology systems related to pay-rolls, attendance, human resources functions, billings, assets management, etc. His mission has been to adopt technology as quickly as possible so that innovation occurs within the institution. With his initiatives to reduce operational costs, a sodium chloride system has been installed, along with the implementation of several energy savings initiatives. Furthermore, in accordance with the pay reform schemes of the Royal Government of Cambodia, a minimum monthly salary for the PPWSA staff has been implemented and increased to improve work efficiency. From HRD perspective, he has managed to "fine-tune" the status of PPWSA in 2014 aiming to straighten the human resource recruitment policy that grants the Director General a special power to select the qualified person from the marketplace.

Sim Sitha's second term as Director General of PPWSA expired in 2018. He has been reappointed as the Director General for the 2018–2021 period.

His first mandate during 2012–2015 was about settling down in PPWSA. His second term during 2015–2018 was used to consolidate operations of PPWSA through technology and HRD. His third term of 2018–2021 is likely to be challenging since the institution will be facing several difficult challenges during the next several years which are likely to be more complex and difficult to solve than what the institution had faced over the past two decades.

Sim Sitha is optimistic on the future of PPWSA. His overall philosophy has been: *Allow a repeat action, present an opportunity. Allow a repeat action* means an opportunity given to staff who fail to do or achieve their tasks, and thus have to redo it properly. *Present an opportunity* means staff are introduced to new ideas which provide them with new opportunities.

Sim Sitha believes that a Director General must have three essential leadership and management qualities to inspire the staff of the institution: technical know-how, accounting and financial knowledge, and human resources management skills (Fig. 3.7).

Fig. 3.7 Sim Sitha

References

Biswas AK, Tortajada C (2009) Water supply of Phnom Penh: a most remarkable transformation. Research Report, Third World Centre for Water Management, Mexico

Chan ES (2009) Bringing safe water to Phnom Penh's City. Int J Water Resour Dev 25(4):597–609. https://doi.org/10.1080/07900620903306323

Colon M, Guérin-Schneider L (2015) The reform of new public management and the creation of public values: compatible processes? An empirical analysis of public water utilities. Int Rev Admin Sci 81(2):264–281. https://doi.org/10.1177/0020852314568837

Curtis G (1993) Transition to what? Cambodia, UNTAC and the peace process. United Nations Research Institute for Social Development, Geneva

Dany V, Visvanathan C, Thanh NC (2000) Evaluation of water supply systems in Phnom Penh City: a review of the present status and future prospects. Int J Water Resour Dev 16(4):677–689. https://doi.org/10.1080/713672527

Das B, Chan ES, Visoth C, Pangare G, Robin S (eds) (2010) Sharing the reforms process. IUCN, Switzerland. https://portals.iucn.org/library/sites/library/files/documents/2010-046_0.pdf

Otis D (2013). How Phnom Penh created a super-efficient, totally drinkable water supply. Fut Resilience, Dec 2. https://nextcity.org/daily/entry/how-phnom-penh-created-a-super-efficient-totally-drinkable-water-supply

PPWSA (Phnom Penh Water Supply Authority) (2009) Clean water for all. Phnom Penh Water Supply Authority, Phnom Penh. https://www.ppwsa.com.kh/Administration/downloads/docs/Brochure1.pdf

PPWSA (Phnom Penh Water Supply Authority) (2017) Phnom Penh Water supply authority third master plan, period 2016–2030, vol 1 (Master Plan Report). Phnom Penh Water Supply Authority, Phnom Penh

PPWSA (Phnom Penh Water Supply Authority) (2020) History. Phnom Penh Water Supply Authority, Phnom Penh

Chapter 4
Four Domain (4-D) Framework for Analysis of Urban Water Utility

Any good and usable framework for analysis has to define its unit of analysis, identify main intended users of the framework analysis and the end objective to be achieved by it. The proposed 4-D framework of analysis is primarily meant for the managers of urban water utilities. They can use it as a management tool to understand the strengths and weaknesses of an individual water utility's business in a comprehensive manner. The end objective of the analysis is to help the urban water utilities to improve its performance from the current state (status quo) to a desired state (aspirational state).

The proposed framework of analysis divides the business of an urban water utility into four fundamental domains: physical, operational, financial and institutional. In all the four domains, a water utility has certain internal forces on which it has control and certain other external forces on which it has very little, or no, control. There are also inter-domain aspects which need proper coordination.

The physical domain analysis provides good knowledge and understanding of demand–supply conditions for the existing situation, and also for the desired state (aspirational state). A water utility may have an aspirational plan to be achieved at a future date. This may include higher and reliable coverage of the population it may serve with piped water supply, provision of higher volume of water on a per capita basis to the population living in its jurisdiction, and supply its consumers with better quality of water. The physical domain analysis is aimed at assessing the ability of the water utility to secure adequate raw water for achieving its future aspirational goals. It also includes calculations of urban water balance (Grimmond and Oke 1986), which may consider variables like precipitation, evapotranspiration, surface run-off and other relevant hydro-climatic factors. Physical domain analysis combines the rigour of accounting with hydrological knowledge and information. The knowledge and information of water cycle and water balance are combined with basic concepts of accounting such as identification, quantification, recognition, presentation and disclosure (Australian Government 2012).

The operational domain analysis includes investigations of all underlying parameters for fulfilling the water supply service obligations to its consumers. It may include analyses of water treatment plant capacities, volume of water production, transmission and distribution pipeline infrastructure details, non-revenue water, customer mix, quality of water, customer care, internal organisational structure, manpower available and their capacities, technology adoption practices and other key and appropriate operational matrices.

The financial domain analysis is aimed at investigating the financial health of a water utility. The analysis could include examination of revenue mix, production costs, unit realisations, tariff structure, financial ratios, etc. Financial domain analysis will provide information on financial resources required by an urban water utility to move from status quo to meet its aspirational goals.

The institutional domain analysis information on institutional capacities for raw water abstractions, raw water quality standards, maintaining drinking water quality standards, economic water regulations, customer care, investment approvals, dispute resolution practices, protection of property rights, etc. The institutional domain in 4-D framework includes only the formal institutions which are designed to perform specific tasks such as raw water abstraction regulation, etc.

The proposed 4-D framework is to certain extent influenced by the work of Saleth and Dinar (2004). In defining the institutional underpinnings of the water sector as a whole, they introduced the concept of physical, economic, policy and institutional dimensions. The physical dimension considers demand–supply analyses of water; economic dimension considers roles of water in socio-economic development such as irrigation; policy dimension covers roles of policy in managing water needs; and institutional dimension examines the roles of institutions that define development, allocation, utilisation, etc. Saleth and Dinar study considered quantitative analysis of cross-country study of institution-performance linkages in the water sector. It is a very comprehensive study across 43 countries, including feedback from 127 water experts on performance-institution linkages in these countries. The 127 respondents were asked to evaluate the water sector's physical performance (demand and supply); operational performance (allocative ease and efficiency); and financial performance (cost recovery and pricing efficiency). The study used an econometric model to assess the roles of institutions on water sector performance.

The 4-D analytical framework proposed in this book has some commonalities but also some significant differences with the Saleth and Dinar process. There is some similarity in nomenclature and underlying meanings of physical dimension and performances described by Saleth and Dinar with the physical domain of the 4-D framework. However, there are significant differences between the underlying meaning of operational performance in Saleth and Dinar's study and operational domain of 4-D framework. The operational performance of Saleth and Dinar considers allocative ease and efficiency in economic terms. In contrast, the operational domain of 4-D encapsulates all the operating activities necessary to deliver good and reliable urban water services. The financial performance of Saleth and Dinar is somewhat similar to the financial domain of the 4-D framework. The institutional domain of the 4-D framework includes only the formal institutions, whereas Saleth and Dinar consider both formal and informal institutions under its institutional domain.

The 4-D framework is aimed at finding the performance gaps in the physical, operational, financial and institutional domains of a water utility. These gaps are often noted in subjective terms rather than in quantitative terms. The 4-D framework does not use any econometric modelling, like ordinary least squares or 3-stage least squares that is normally considered in studies of institutional economics of water, as in the Saleth and Dinar's study. The 4-D framework is subjective, intuitive and comparatively simple to be understood by an urban utility manager of a developing country city. A typical value chain of using data to final impact is shown in Fig. 4.1. The proposed 4-D framework segments analysis into four distinct domains. Though the proposed 4-D framework recognises the interplay of variables between the four domains, the 4-D framework has limitations as it does not provide explicit ways to make subjective or quantitative connections between the four domains.

It should be noted that a perfect framework for analysis of an urban water utility may be desirable but may neither be achievable nor essential under most circumstances. Accordingly, the proposed 4-D framework should be considered as a design thinking concept. There have been positive feedbacks from several water utilities about the simplicity of the 4-D framework and its ability to cover most of the business aspects of running an urban water utility efficiently.

PPWSA is analysed in this book through the lens of the proposed 4-D framework. It should be noted that this proposed 4D framework is a work in progress. As feedback is received from various water utility managers, donor agencies, academics, practitioners and public policy experts, the framework will improve progressively. After the next iteration of an improved framework, based on new feedback received, it will be used by the authors to analyse 10–15 Asian water utilities in the coming years.

Fig. 4.1 Value chain of 4-D framework

References

Australian Government (2012) The national water account companion guide. Commonwealth of Australia, Melbourne. https://www.bom.gov.au/water/nwa/document/companion-guide.pdf

Grimmond C, Oke TR (1986) Evapotranspiration rates in urban areas. IAHS-AISH Publication 259

Saleth RM, Dinar A (2004) The institutional economics of water: a cross-country analysis of institutions and performance. World Bank Washington, DC. https://documents1.worldbank.org/curated/en/782011468780549996/pdf/302620PAPER0In1l0economics0of0water.pdf

Chapter 5
Analysis of PPWSA

The Four-Domain (4-D) Framework is used in this section to analyse PPWSA "business" in its four key domains.

5.1 Physical Domain

This section analyses the quantity and quality of raw water available for Phnom Penh and PPWSA to meet its obligations to supply drinking water to an increasingly larger population. It also considers the quantity of raw water required by PPWSA to meet the water demand of Phnom Penh by 2030.

5.1.1 Rivers and Seasonal Direction of Flow

Phnom Penh Municipality is located to the west of a X-shaped confluence (called "Chaktomuk") of the Mekong, Tonlé Sap and Bassac rivers. Raw water for Phnom Penh is abstracted from this confluence (PPWSA 2017a). A schematic of the location of Phnom Penh with respect to the confluence of the rivers is shown in Fig. 5.1.

Mekong is the 12th longest river in the world. It runs 4800 km through Tibet in China, Myanmar, Thailand, Cambodia, Lao PDR and Vietnam. The river has an average flow of 17,000 m^3/s (PPWSA 2017b).

Sap River connects Mekong to the Tonlé Sap Lake located about 150 km away in the central plains of Cambodia. During the early rainy season (June–August), when the Mekong level tends to rise, this river flows into the Tonlé Sap and eventually into the big lake. After the early rainy season, the flow direction is reversed. The lake then flows into the Tonlé Sap river and then eventually into the Mekong river. Tonlé Sap has lower flows during the dry season (around 500 m^3/sec) (PPWSA 2017b).

© The Author(s), under exclusive license to Springer Nature Singapore Pte Ltd. 2021   49
A. K. Biswas et al., *Phnom Penh Water Story*, Water Resources Development
and Management, https://doi.org/10.1007/978-981-33-4065-7_5

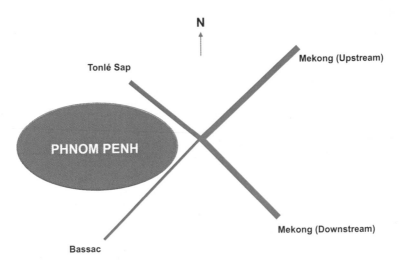

Fig. 5.1 Water abstraction for Phnom Penh

Bassac River is relatively a small river and it flows in the southern part of Phnom Penh. It has very limited flow in the dry season, about 40 m³/s in April (PPWSA 2017b).

In the south Phnom Penh, Boeung Tompun acts as the natural outflow of rainwater and most sewage generated in the city. Boeung Tompun is getting progressively backfilled (PPWSA 2017b).

During the early rainy season from June–August time period, 100% of the surface water resource in the area comes from the Mekong River. Sap River flows away from Mekong as shown in Figs. 5.2 and 5.3.

Tonlé Sap flow inversion takes place during September–November period, when it flows into the Mekong. However, Mekong has higher velocity and Tonlé Sap river has a lower velocity. The water downstream of Phnom Penh is thus a mixture of Mekong upstream and the Tonlé Sap river.

After rainy seasons, during November-February, the flow direction remains the same as the September–November period. The only difference is that during this period Mekong river's flow and also its velocity is lower. Much of the water downstream of the right bank of Phnom Penh comes from the Tonlé Sap river.

During dry seasons, the flow in the Sap river decreases close to zero, However, flows in the Mekong remain somewhat similar. Hence, the influence of the Sap river in the downstream of Chaktomuk becomes less important.

Fig. 5.2 Early rainy season
river flow (June–August)

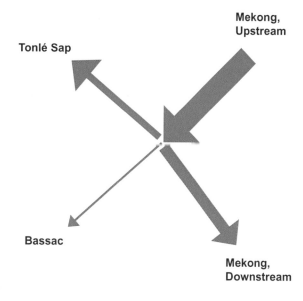

Fig. 5.3 Tonlé Sap inversion
(September–November)

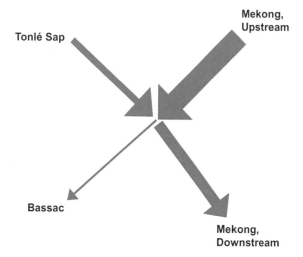

5.1.2   Water Treatment Plants, Their Location and Source of Surface Water

Phnom Penh has four water treatment plants (WTP) at Chroy Changvar, Phum Prek, Chamkar Mon and Niroth. The raw water source for the four water treatment plants is shown in Fig. 5.4.

Phum Prek WTP faces water quality issues during the summer months. As a natural phenomenon, when water is relatively stationary like in case of a reservoir

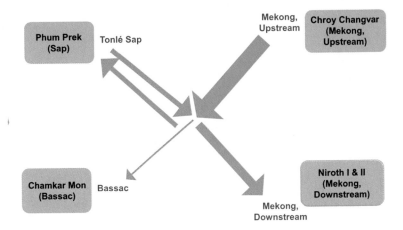

Fig. 5.4 Location of the Water Treatment Plants and their source of water

or water has slow movements due to low river flows, the water surface becomes warmer. Warm temperature, plenty of sunlight, and good sources of nutrients are factors that encourage algae growths. During the summer, the water of the Tonlé Sap river becomes almost stagnant, that is no flow for all practical purposes. This leads to algae growth and thus high eutrophication rates. Algae reaches the Phum Prek water treatment plant through intake pumps. Most of the algae are difficult to settle and thus they could clog the filters. Phum Prek WTP uses pre-chlorination during summer months to avoid problems due to algae presence in raw water.

5.1.3   Phnom Penh's Raw Water Requirement and Prime Source of Raw Water

There is limited groundwater potential in Phnom Penh. Accordingly, the entire water demand of raw water for Phnom Penh has to be met by surface sources (PPWSA 2017b).

Among the surface water sources, Mekong river has an average flow rate of 17,000 m^3 per second. Its monthly average flow at Chroy Changvar station is shown in Fig. 5.5. The dry season flow of Sap river is only 500 m^3 per second and the flow of Bassac is even less, only 50 m^3 per second.

In 2016 PPWSA extracted 0.22% of the annual minimum Mekong flow (about 500 MLD), that is 0.04% of the average annual Mekong flow, and 0.01% of its maximum annual flow (PPWSA 2017b).

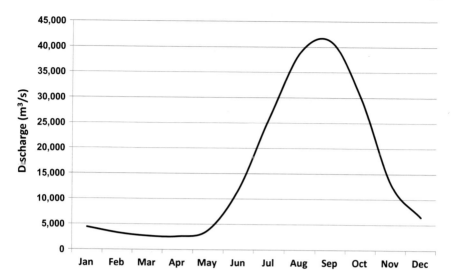

Fig. 5.5 Monthly average Mekong Flow at Chroy Changvar Station (1960–2014). *Source* PPWSA (2017b)

Over the long run, dams, whose construction are being planned on the Mekong by Cambodia as well as other countries upstream of Cambodia, and also on its main tributaries, may reduce the current flows of the Mekong from which Phnom Penh can abstract water. As of now, several dams are under construction in China and around 12 mainstream dams have been proposed in China, Lao People's Democratic Republic, Thailand and Cambodia. The two dams planned for Cambodia are located at Sambor and Stung Treng, upstream of Phnom Penh. Both the dams are primarily for hydropower generation that is expected to generate electricity for an increasingly energy-stored Cambodia to ensure its future economic and social development.

In early 2020, Cambodia started to import electricity from its neighbour, Laos, from the new Don Sahong hydropower complex, having an installed capacity of 260 MW, under a 30-year agreement. Following this agreement, Cambodia has announced no new dam will be constructed on the Mekong before 2030. Sambor Dam, when constructed, will be the smallest dam on the mainstream Mekong River. However, still this will be the largest dam in Cambodia if and when it is constructed.

5.1.4 Future Raw Water Demand and Location of New Water Treatment Plants

The Third Master Plan (2016–2030) of PPWSA made detailed estimates of water demands to 2030. The prime variables used for making demand projections are growth in number of water connections, density of connections (number of connections per hectare), per capita daily domestic consumption, commercial and industrial water requirements and expected non-revenue water by 2030.

The number of connections is projected to increase from 300,000 in 2016 to 550,000 in 2030 implying a compounded annual growth rate of 4.40% during this period. The average daily demand in 2016 is expected to grow from 488,827 to 950,478 m^3/day by 2030. Similarly, the maximum daily demand is expected to grow from 562,151 in 2016 to 1,093,050 m^3 by 2030. It is expected that additional capacity of 400 to 500 MLD will be required by 2030. Under this plan, Phnom Penh will need new water treatment plants with a combined capacity of 400–500 MLD by 2030.

Phnom Penh has to consider two important issues. These are: where to get the raw water from, and where to locate the new water treatment plants?

Bassac river is a small river with dry season flows of less than 40 m^3/second. There are two small new water treatment plants planned on this river. With these new facilities, 3% of the Bassac river flow will be used by all the existing and planned water treatment plants. Thus, this river has limited capacity to provide an incremental source of raw water in the future for Phnom Penh city.

The two main sources of raw water for Phnom Penh could be the Tonlé Sap and the Mekong rivers. Tonlé Sap's dry season flow is only 500 m^3/s. Furthermore, the water quality of this river depends on untreated wastewater that is discharged from the Phnom Penh city, much of which eventually finds its way to the Tonlé Sap river. In addition, its water quality is also susceptible to quality issues due to change in direction of its flow. In contrast to the Tonlé Sap, the Mekong river does not have water quality issues or flow constraints. A total of 1000 MLD water needs of Phnom Penh, in 2030, can be met by extracting about 0.44% of the annual minimum flow, 0.08% of the average annual flow and 0.03% of the maximum annual flow. Thus, Phnom Penh is well-placed to abstract its future extra water requirements from the Mekong river.

The Mekong river in Phnom Penh has two segments, Mekong upstream and Mekong downstream. Existing Niroth water treatment plants (Phase I and Phase II) contribute 44% of the total water treatment capacities. They are both based on the Mekong downstream. Thus, a suitable location for a new water treatment plant for Phnom Penh could be from the upstream of this river.

5.1.5 Outlook on Raw Water for Phnom Penh

The Mekong river has sufficient flows to support the growing water demands of Phnom Penh. It can support the needs of raw water demands of Phnom Penh till at least 2030.

However, an important challenge for Phnom Penh is the absence of wastewater treatment capacities. The terrain slope of the Phnom Penh is oriented from north-west to the south-east, and elevations range from 15 to 5 m above sea level. Southern part of the city acts as the natural outlet of rainwater and most sewerage water generated in the city. However, the *boeungs* (lakes) that act as natural cleaning mechanisms for wastewater are fast disappearing due to new property and road developments. The challenges related to wastewater discharge are discussed later in this book.

5.2 Operational Domain

This section discusses PPWSA's water supply system's components, including capacity of water treatment plants, production by water treatment plants, distribution pipeline network, area of coverage, timings of water supply, non-revenue water, consumer mix, water quality aspects, consumer complaints, internal organisation structure and key manpower resources available at PPWSA. PPWSA datasets have been used to analyse the aforesaid components in this section. The strengths and the weaknesses of the operational domain are also noted.

5.2.1 Capacity and Production of Treated Water

The four water treatment plants of Phnom Penh are located at Chroy Changvar, Chamkar Mon, Phum Prek and Niroth. As of June 2017, PPWSA had a total water treatment capacity of 560,000 m^3/day as shown in Table 5.1.

Total water production capacity in 1993 was 65,000 m^3/day. In 1995, the Chamkar Mon water treatment plants (additional 10,000 m^3/day) and the Phum Prek water treatment plant (additional 45,000 m^3/day) were rehabilitated. In 1997, capacity of the Chamkar Mon water treatment plant was further expanded by 10,000 m^3/day. The capacities of the Chroy Changvar (stage 1) water treatment plant was increased by 65,000 m^3/day in 2002. Similarly, in 2013, the capacity of the Phum Prek water treatment plant was augmented by 50,000 m^3/day. The available capacity for water production and actual production are shown graphically in Fig. 5.6.

Table 5.1 Details of water treatment plants, June 2017

Location	Year activated	Year rehabilitated or expanded	Capacity (m^3/day) (June 2017)
Chroy Changvar	1895	2003 and 2009	130,000
Chamkar Mon	1958	1995 and 1997	20,000
Phum Prek	1966	1995 and 2003	150,000
Niroth	2013 (Stage I)		130,000
	2017 (Stage II)		130,000

Source PPWSA

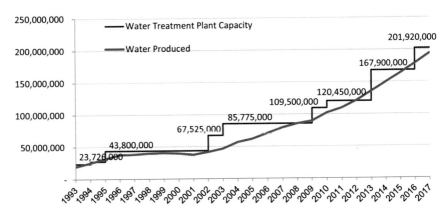

Fig. 5.6 Capacity and production of treated water (m³/year), 1993–2017. *Source* PPWSA

In 2009, the capacity of the Chroy Changvar plant (stage 2) was increased by 65,000 m³/day. The Niroth Stage I plant was commissioned in 2013, having a production capacity of 130,000 m³/day. The Niroth Stage II plant, with a capacity of 130,000 m³/day, became operational in 2017. Currently, the total water production capacity of PPWSA stands at 590,000 m³/day, which is nearly nine times of its capacity in 1993. This expansion, in about 24 years, by any account, has been a most creditable achievement.

Figure 5.6 indicates that it has taken 3–4 years to fully utilise all its additional new water production capacities. This indicates strong underlying latent demand for water in Phnom Penh.

5.2.2   Distribution of Pipeline Length

Total distribution and transmission pipeline length of PPWSA has increased nine times during the 1993–2016 period. This is consistent with the increase in water treatment production capacity which increased by nine times during the 1993–2016 period. Figure 5.7 shows year-wise increase in transmission and distribution length. The water supply system of PPWSA consists of a single pressure network of above 300 mm diameter having a length of 220 km, four water towers each having capacity of 1500 m³, and a distribution network of 2240 km (below 250 mm diameter) (PPWSA 2017b). The water supply system is shown in Fig. 5.8.

As of 2016, PPWSA piped water supply area covered 229 km², which is 32% of the total Phnom Penh municipal area. Before the addition of 20 communes from Kandal Province into Phnom Penh in 2010, the total area of Phnom Penh municipality was 376.2 km². This implies that the current piped water supply area covered 61% of Phnom Penh Municipal area before 2010 (PPWSA 2017b).

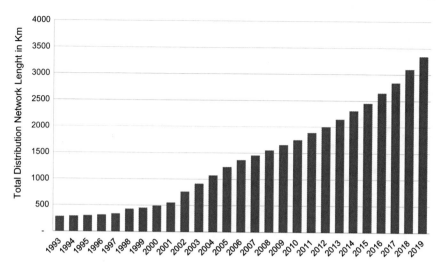

Fig. 5.7 Distribution network lengths in km, 1993–2019. *Source* PPWSA

The total area PPWSA has been supplying water has increased significantly during the 2005–2015 period. This is shown in Fig. 5.9. This increase in area is one of the reasons that the production of water has increased beyond the level of water treatment plant capacity in 2015 as shown in Fig. 5.6.

PPWSA uses ductile iron and high-density polyethylene pipes for water transmission and distribution. Ductile iron pipes are used exclusively for diameters of more than 250 mm. High-density polyethylene pipes have been used since 1996, and thus far no problems have been observed. However, for these types of pipes special attention has to be paid to ensure that the joints are properly done so that there are no leakages when such pipes are joined with other similar pipes or pipes made of other materials. This requires good electro-fusion, especially for pipes made of other materials. On the cost side, a 50–500 mm diameter pipeline typically costs $40–70 per meter for excavation and backfilling. For sizes, greater than 50–500 mm, the cost depends on the design considerations. A 1500–2000 m transmission line can cost $300–500 per meter.

5.2.3 Performance of the Water Supply Network

Phnom Penh consumers continue to receive a continuous and reliable 24-h water supply. However, discussions with certain staff members and some customers indicated that there are certain areas in Phnom Penh where consumers in second and higher floors of buildings are witnessing low water pressures during the peak hours of water consumption, especially during 6.00–8.00 am in the morning. The lack of adequate water pressure in some specific areas of the city is due to the rapid expansion

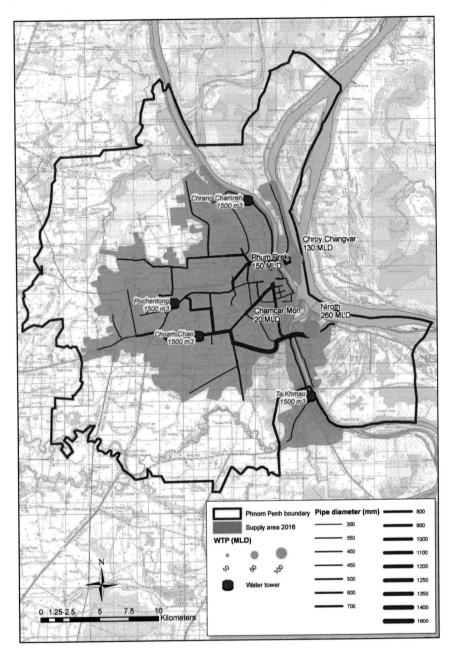

Fig. 5.8 PPWSA Water Supply System. *Source* PPWSA (2017b)

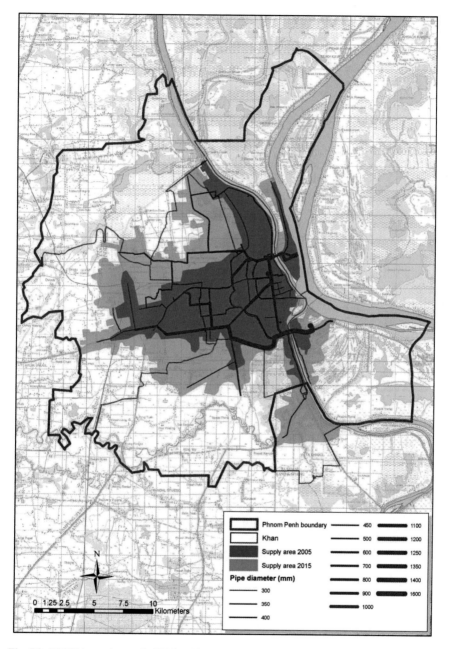

Fig. 5.9 PPWSA supply area in 2005 and 2015. *Source* PPWSA (2017b)

of the water supply network to peripheral areas. This expansion of the distribution network was beyond the original hydraulic design of the water transmission network. The PPWSA Master Plan document (2016–2030) notes "the lack of pressure remains the major obstacle to proper implementation of zoning."

5.2.4 Non-revenue Water and Unaccounted for Water

PPWSA established a water loss control committee in March 2003, with responsibility to monitor, evaluate and review the water loss as well as take necessary measures to keep water loss below 10%. Members of the committee are key staff from relevant departments. Furthermore, an official water balance table adapted to the IWA water balance methodology was introduced since 2014.

Non-revenue water (NRW) and unaccounted for water (UFW) are amongst the key indicators of operational performance of any water utility. Non-revenue water for a water utility is generally estimated as follows:

$$\text{NRW (in \% terms)} = (1 - (WV_{BILLED)}/(WV_{PRODCUED)})) * 100\%$$

where

WV_{BILLED} Volume of Water sold and billed by the water utility, and.
$WV_{PRODUCED}$ Volume of water produced by the water utility.

NRW's different constituents for any water utility are shown in Fig. 5.10. The sum of apparent and real loss components shown in Fig. 5.10 is called unaccounted for water.

In 1993, non-revenue water of PPWSA was extraordinarily high, at 72%. With good management, this was reduced to 5.9% by 2009. A detailed discussion about the reasons behind this impressive reduction has been discussed earlier. One of the main reasons for this significant reduction from 72 to 5.9% in only about a decade was that the employees were incentivised to reduce non-revenue water as much as possible. Improvements in the underlying pipeline network also contributed to the reduction of non-revenue water. However, after 2009, NRW has been inching up.

Figure 5.11 shows that the non-revenue water has increased from 5.85% in 2012 to 8.16% in 2019. This increase can be partly explained by the increase in the age of the distribution pipeline in recent years as shown in Fig. 5.12. The age of the pipeline network was the lowest in 2006. Since then the average age has been gradually increasing. This could be another possible reason for the gradual, but steady increase, of non-revenue water from 2009 onwards. Higher NRW could potentially adversely affect the profitability of the utility in the coming years unless appropriate countermeasures are taken in a timely manner (Fig. 5.13).

This performance in NRW is significantly better than Thames Water of the UK which was fully privatised in 1983. It was fined 8.5 million by the water regulator,

System input volume	Authorised consumption	Billed authorised consumption	Billed metered consumption (including water exported)	Revenue water
			Billed non-metered consumption	
		Unbilled authorised consumption	Unbilled metered consumption	
			Unbilled non-metered consumption	
	Water losses	Apparent losses	Unauthorised consumption	Non-revenue water
			Metering inaccuracies	
		Real losses	Leakage on transmission and/or distribution mains	
			Leakage and overflows at utility's storage tanks	
			Leakage on service connections up to customers' meters	

Fig. 5.10 NRW Constituents. *Source* Liemberger (2002)

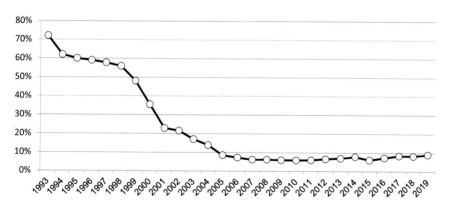

Fig. 5.11 Non-revenue water, 1993–2019. *Source* PPWSA

OFWAT, in 2017, the maximum amount permissible under the regulation. This private sector company lost 35 million litres per day, equivalent to 180 L per day per property. Its losses actually rose 5% in 2016 compared to 2015. In contrast, PPWSA, a water utility in a developing country lost, as a percentage of total production, only about 1/3rd of Thames Water. On NRW, PPWSA's performance has been significantly better than many big cities of Europe and North America.

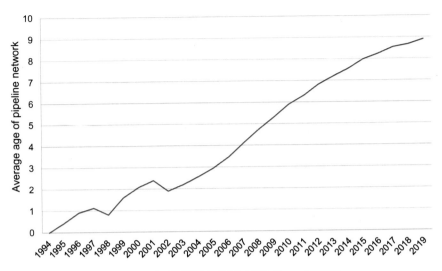

Fig. 5.12   Age of pipeline network of PPWSA, 1995–2019. *Source* PPWSA

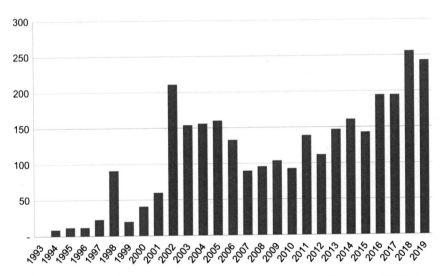

Fig. 5.13   Increase in distribution network length every year (m), 1993–2019. *Source* PPWSA

Furthermore, lengths of the distribution network started to increase significantly and steadily every year, from 2002 onwards. This rapid increase may also have partially contributed to the increase in NRW.

5.2.5 Water Supply and Sanitation Branch

The Water Supply and Sanitation Branch of PPWSA is responsible for all the water pipeline construction activities that the utility undertakes for itself as well as for external clients. This Branch lays 150–200 km of pipeline every year. Cambodia did not have adequate expertise in laying pipes during the 1990s. Thus, PPWSA had to develop the necessary expertise in-house for water infrastructure construction. PPWSA can lay the pipelines cheaper and the quality of its construction is significantly better than its private counterparts.

This Branch has a total staff of 156 which is responsible for water infrastructure construction. After listing of PPWSA on the Cambodia Stock Exchange, a separate subsidiary was created to be responsible for the construction activities within the company as well as provide its services to other provinces to lay pipes. The current internal capacity of PPWSA in construction planning and management is excellent. This brings many financial benefits to PPWSA. However, its main challenge in executing the construction activities for other Cambodian provinces is that the human capacity of counterpart institutions in the provinces continues to be rather low. By constantly finding implementable solutions for executing projects in other provinces, PPWSA has been continually upgrading its own project management skills.

5.2.6 Customer Mix

PPWSA has primarily three kinds of customers, that is, domestic, commercial and government. Domestic customers account for 87% of total connections in 2002, and 81% of total connections in 2016. Commercial connections further accounted for 17% of total connections in 2016. However, in terms of volume consumed in 2016, the domestic consumers were responsible for 51% of total volume of water consumed, and the commercial customers accounted for another 40% in 2016. The same year, domestic and commercial customers contributed 45% each to the total revenue of PPWSA. The mix of customers in terms of connections, volumes sold and value sold for the 2000–2019 period are shown in Figs. 5.14, 5.15 and 5.16 respectively.

There were 13,634 domestic connections, or 2.2% of the total domestic connections, in December 2017, that consumed no water. The household connections that consume less than 7 m^3 of water per month are 37.7% of the total number of domestic connections. Altogether they consumed 23% of the total volume of water used by the domestic sector. The average monthly domestic consumption, based on total number of domestic connections in December 2016, was 27.53 m^3.

There is a segment of customers in Phnom Penh that is unique to PPWSA. It is called "rented rooms." These are houses which are connected to PPWSA's water pipeline network and these households rent rooms to others. Landlords of such houses buy water from PPWSA and then resell the water to the tenants through metered

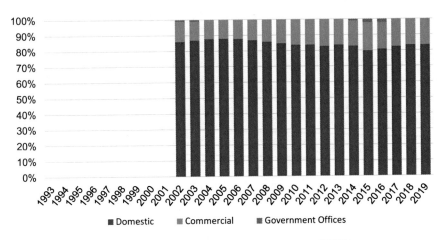

Fig. 5.14 Consumer mix in terms of connections, 1993–2019. *Source* PPWSA

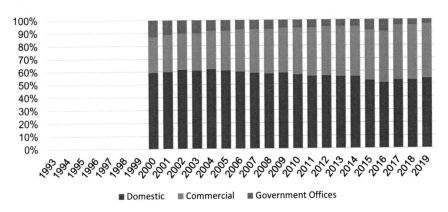

Fig. 5.15 Consumer mix in terms of volumes, 1993–2019. *Source* PPWSA

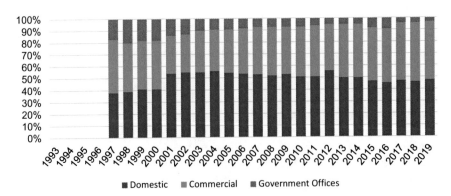

Fig. 5.16 Consumer mix in terms of values, 1993–2019. *Source* PPWSA

connections to each rented room. The households that sell water to rented rooms are required to disclose the information on the number of rooms rented in the houses to PPWSA. PPWSA sells water to these households through a metered connection. These households, in turn, sell water to the rented rooms through individual meters. As of July 2017, PPWSA had 5571 "rented room" segment connections, which in turn had 121,313 rented rooms attached to them. The "rented room" segment consumes 5–6% of the total water volume sold and billed by PPWSA. It is estimated that 2–8 people stay in each of the rented rooms. The people staying in the rented room are mostly migrant workers, who are employed in the garment and other factories in Phnom Penh. The landlord installs separate meters for all individual rooms.

Landlords of the "rented room" segment are advised by PPWSA not to charge more than KHR 100/m^3 of premium to the tenants of the rented rooms, above the tariff charged by PPWSA to the landlords. It is highly likely that in practice the landlords of the rented rooms are charging a premium of more than KHR 100/m^3 to the tenants to cover the costs of installing meters or pumps, and perhaps also to make higher profits.

5.2.7 Quality of Water

Every year, PPWSA sends water to oversee independent laboratories like Singapore and Shanghai, China, for testing over 100 parameters.

The quality of water supplied by PPWSA complies with the Cambodian National Drinking Water Quality Standard (CNDWQS) as prescribed by the Ministry of Industry and Handicrafts (MIH). As per this Standard, 47 parameters are to be tested for water quality, of which 6 parameters are to be tested daily, 16 are to be tested every three months, 26 are to be tested once a year and the rest of the 47 parameters are to be tested every three years. PPWSA is required to provide water quality samples to MIH every three months. PPWSA, on its own volition, does water quality tests on more number of parameters every day, weekly, monthly and on yearly basis compared to what is mandated by the MIH. PPWSA collects 80 samples from its service area every week for water quality tests. At present PPWSA does not publish any annual water quality report.

5.2.8 Customer Care

In the final analysis, urban water supply is a "service" business and not a "product" business. In a "product" business, a one-time sale is made by the company to a customer but in "service" business the company is in regular contact with its consumers. Though the customers are important in both "product" and "service" businesses, the companies in the "service" business must make extra efforts to keep their customers regularly engaged. Many water utilities still do not realise that they are in the "service" business and not in a "product" business.

In the case of PPWSA, consumers can complain about their problems to PPWSA in person, by telephone, or through Facebook. There is no provision for online filing of complaints by the consumers. The call centre supporting the complaints made through telephone is managed by 2–3 PPWSA employees. Currently PPWSA receives about 1000 complaints every month. Of these, 200 complaints are received via PPWSA's Facebook account. The mix of customer complaints currently being received is shown in Fig. 5.17.

During our interviews with twenty customers of PPWSA, some of the interviewees mentioned that they were not aware about the best way to contact PPWSA about their complaints. There is a need for PPWSA to make its customers more aware of the best way to contact the utility with their views and complaints. It may be necessary to deploy more personnel in their call centres to handle an increasing number of calls promptly and satisfactorily. As the internet connections in Cambodia have increased exponentially in recent years, the utility may have to make a provision to accept consumer complaints online in the near future. It could also install a software that can track the nature of the complaints and also time taken for the resolution of the complaints. Such tracking of the complaints and regular assessments of how they

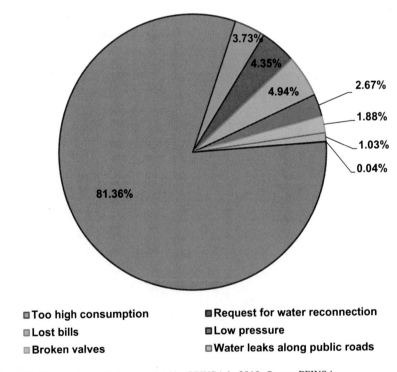

Fig. 5.17 Nature of complaints received by PPWSA in 2019. *Source* PPWSA

were handled would undoubtedly improve both the performance of the utility and customer satisfaction rates.

5.2.9 Organisation Structure and Management Team

Organisationally PPWSA is divided into six departments. The organisational structure and the six departments of PPWSA are shown diagrammatically in Fig. 5.18.

PPWSA is governed by a Board of Directors with 7 members and a 3-year mandate. The Board is headed by a representative from MIH. Another 3 members are nominated by the Government too, a representative from MEF, a representative from Phnom Penh Municipality, and the Director General. The other 3 elected members are: a representative of employees elected by PPWSA's employees; an independent director elected by all shareholders; and a non-executive director representing private shareholders elected by private shareholders.

Currently PPWSA has a very strong and experienced top management team. It has seven Deputy Director Generals. Dr. Chea Visoth heads the corporate secretariat that reports directly to the Director General. Dr. Visoth has also been the Deputy Director General for Production and Water Supply since November 2017. He joined PPWSA in 1994, and has a Ph.D. degree from Germany. Like many of his peers, he had an opportunity to study abroad and has very good knowledge and skills for the planning and management of water utilities. He was earlier the head of the training programme, an important function that has significantly contributed to the success of PPWSA.

Ros Kimleang is a Deputy Director General responsible for the Accounting and Finance Department. He joined PPWSA in 1980 as a field worker. He studied in Russia from 1982 to 1985. Thereafter, he worked as a technician. He studied financial accounting from Institute of Finance and Accounting in Phnom Penh during 1987–1991. He further has a Master's degree in finance from study abroad during 2001–2003. He is a humble but very competent person and is very passionate about ensuring continued success of PPWSA (Fig. 5.19).

Samreth Sovithiea is Deputy Director General and the head of the Planning and Project Department. He joined PPWSA in 1992, immediately after his graduation. He became a Chief Officer and then promoted from middle to senior management. He brings to PPWSA a very rich experience due to his ability to interact with various external agencies, including different ministries and donor organisations. The role of the Planning Department is a very critical one. One of its many responsibilities is the preparation of the Master Plans in collaboration with the donors. The Three Master Plans, the first of 1993–2010, the second of 2006–2020, and updates of 2008 and 2012, and the third plan of 2016–2030 are the backbone of the planning process.

Long Naro is Deputy Director General and responsible for water supply and sanitation. He is also the head of the NRW monitoring committee that is responsible for keeping NRW low. Long Naro completed his graduate degree in mechanical engineering in 1990 from Germany. He initially worked with an NGO, Oxfam, that

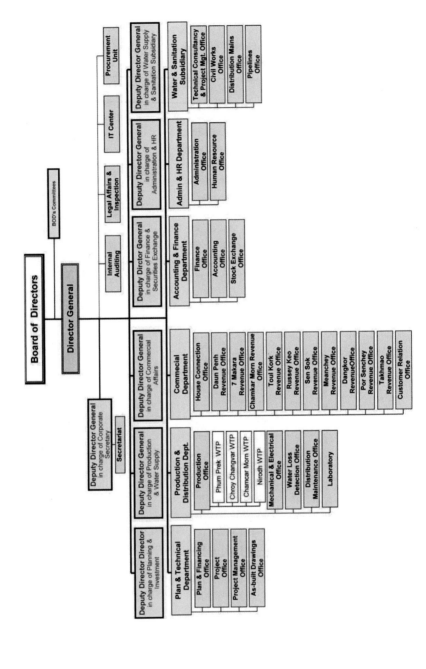

Fig. 5.18 Organisational structure. *Source* PPWSA (2017a)

Fig. 5.19 Chea Visoth and
Ros Kimleang

Chea Visoth   Ros Kimleang

was helping to rehabilitate the Phum Prek water treatment plant. In 1992, Long Naro joined PPWSA as Deputy Chief of the water treatment plant. He worked with JICA on the first master plan. In 1993, Long Naro became Chief of the Design and Technology Office. In 1997, he became the Chief of Design and Technical Department. In 2002, he assumed the role of Deputy Director General for design, management of water resources. From 1996, he has also been supervising all the construction activities of PPWSA.

Ros Kimleang, Samreth Sovithiea and Long Naro have also a degree in business management from Charles Sturt University, of Australia, where they studied during 2000–2003, with the financial support from the World Bank (Fig. 5.20).

From this description, it is evident that PPWSA has a very strong top management team which has been an integral part of the journey of its turnaround from the early 1990s. What is noteworthy is that this management talent is "homegrown." The strong work ethics of PPWSA management team, their firm commitments to its success, and their long stay with the utility, have been three main reasons for the continued excellence of PPWSA's planning, management and operational performances.

Fig. 5.20 Samreth Sovithiea
and Long Naro

Samreth Sovithiea   Long Naro

Fig. 5.21 Ngin Chantrea
and Chan Piseth

Ngin Chantrea   Chan Piseth

Amongst the other current senior members of the management team of PPWSA are Ngin Chantrea who has been Deputy Director General for Commercial Affairs of PPWSA since 2015 and Chan Piseth, Deputy Director General for Human Resources since 2014. Ngin Chantrea joined PPWSA in 1993 and she worked in the Finance and Accounts Department from 1993 to 2007. In 2008, she joined the Commercial Department. In 2015, she became the Deputy Director General of the Commercial Department. Her key focus has been to maintain the collection ratio high, improve the standard operating procedures for her department and maintain the integrity and discipline amongst its meter reading staff. Chan Piseth joined PPWSA in 2014 from the Ministry of Industry and Handicrafts. His focus has been to improve the standard operating procedures of the Human Resources Department and to improve the training need assessment planning, which includes both annual and over the long-term (Fig. 5.21).

5.2.9.1 Outlook on Operational Performance

PPWSA has done an absolutely remarkable job in terms of reducing NRW from 72% in 1993 to less than 8.94% by 2019. During the 2009–2011 period, the NRW was below 6%. During the same period, its network of pipelines expanded by over nine times. This is a remarkable performance of NRW reduction which is unparalleled in the entire history of water utilities from any city of any developing country.

The key operational challenge PPWSA faces is expansion of Phnom Penh city in terms of geographical area and population. PPWSA needs to carefully consider the expansion of its water distribution network beyond the hydraulic capacity of its current water distribution network. For water utilities in all developing countries, there are often strong and sustained political and social pressures to continually expand its pipeline distribution networks to newer and newer areas. However, such expansions may compromise the hydraulic integrity of the network. If and when the hydraulic capacity of the network is compromised, it invariably leads to lower pressures in certain areas during peak consumption hours as the first symptom of the weakness of the water distribution network. Thereafter, if the expansion continues, it could lead to an eventual situation where 24-h water supply itself may be compromised.

In addition, PPWSA also needs to guard itself against higher NRW as the average age of its pipeline network continues to increase. This is already becoming evident. NRW has started to increase gradually from 5.95% in 2011 to 8.94% in 2019.

An expanding Cambodian economy will not only create new employment and economic opportunities for the Cambodian people, but it could also put pressure on higher salary expectations in the job market. Accordingly, in the coming years, PPWSA may face significant challenges in hiring talented manpower against stiff competition for good talents from other employers, and also retaining them over the long-term.

5.3 Financial Domain

This section provides an independent analysis of the financial performance of PPWSA from 2008 onwards. This assessment is based on the analyses of independent audits of the accounts of PPWSA during this period. The utility's accounts were audited by PricewaterhouseCoopers from 2011 to 2014, BDO (Cambodia) in 2015 and Grant Thornton in 2016. The frequent changes in auditing firms for PPWSA can be explained by a regulation in Cambodia that mandates that if the cost of the annual audit is more than USD 7000 per year, then the company requesting audits must call for bids to carry out the audit. The cost of annual audit for PPWSA is more than USD 20,000. Hence, there has been frequent changes in the auditing companies that have carried out independent audits of PPWSA as per the national legal requirements.

PPWSA has consistently made net annual profits during the 1998–2019 period (Fig. 5.22). Its water sales, between 2008 and 2016, grew at a compounded annual rate of 10.52%. Its earnings before interest, depreciation and tax (EBITDA) have grown at a compounded annual rate of 8.06% during the same period. Lower compounded growth of EBITDA, as compared to revenue growth between 2008–2016, probably signifies that the cost pressures were building up for the utility. This is not surprising since the water tariffs in Phnom Penh remained unchanged from 2001 to 2016.

5.3.1 Revenue Mix

The revenue mix during the 2008–2017 period is given in Fig. 5.23. Not surprisingly, the dominant share of the revenue came from its water sales. The fee from construction services comes from its in-house construction department called Water and Sanitation Branch. The construction service fees are volatile in nature.

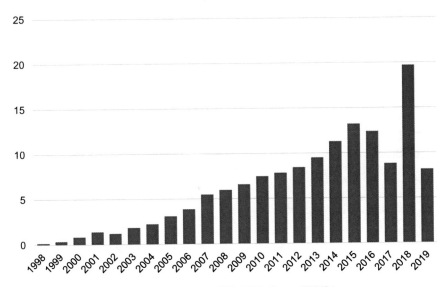

Fig. 5.22 Net annual profits in USD million, 1998–2019. *Source* PPWSA

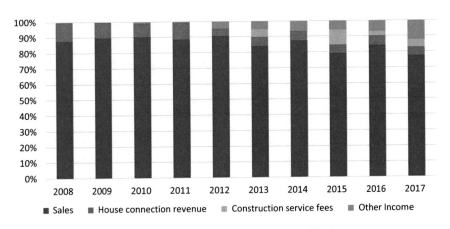

Fig. 5.23 Revenue composition of PPWSA, 2008–2017. *Source* PPWSA

5.3.2   Production Cost

The unit production cost of water, including depreciation and finance cost, has gone up from KHR 724/m^3 in 2008 to KHR/m^3 794 in 2019. The operating cost of water (excluding depreciation) has increased from KHR 483/m^3 in 2008 to KHR/m^3 5187 in 2019 (Fig. 5.24).

There has been a steep increase in interest cost from KHR 15/m^3 of water produced in 2008 to KHR/m^3 52 in 2019.

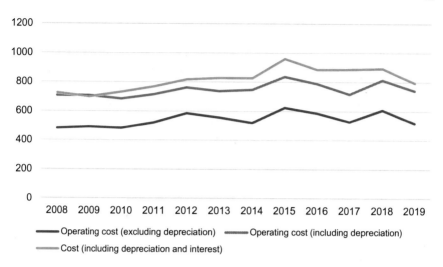

Fig. 5.24 Unit production costs (KHR/m^3), 2008–2019. *Source* PPWSA

The cost composition, excluding depreciation and interest costs, is shown in Fig. 5.25. The electricity cost, as a percentage of the total cost operating cost, has come down from 43% in 2008 to 32% in 2019. However, the employee costs have gone up from 31% of total cost in 2008 to 45% by 2019 (Fig. 5.25).

The unit electricity cost for every m^3 of water produced came down from KHR/m^3 206 of water produced in 2008 to KHR/m^3 168 in 2019, a decrease of 18%. This has been helped by unit electricity cost which fell from KHR 897/kWh in 2008 to KHR

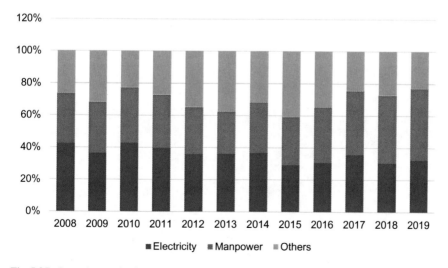

Fig. 5.25 Operating production cost mix (excluding depreciation and interest), 2008–2019. *Source* PPWSA

704/kWh in 2016, a fall of 21.5%. Thus, the reduction in electricity charges per m³ from 2008 to 2019, is primarily due to reduction in unit cost of electricity.

The unit employee cost for every m³ of water produced went up from KHR/m³ 151 in 2008 to KHR/m³ 232 in 2019, an increase of 54%. (Fig. 5.26).

The number of employees per 1000 connections are shown in Fig. 5.27. Total number of employees reduced from 591 in 1993 to 523 in 2003. However, during the same period 1993–2003, the number of employees per 1000 connections saw a steep decline from 22.0 to 4.9 implying a magnificent improvement in employee productivity. The number of employees per 1000 connections reached its lowest at

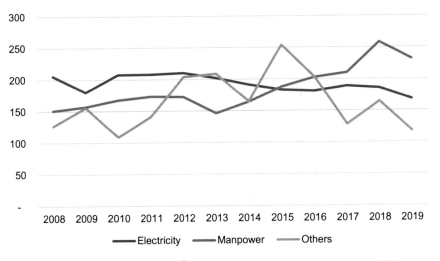

Fig. 5.26 Unit production cost (KHR/m³ of water produced), 2008–2019. *Source* PPWSA

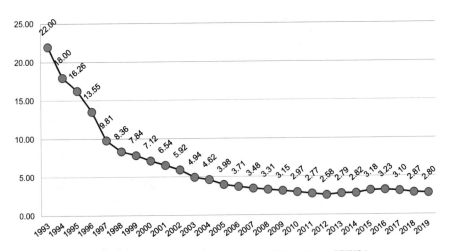

Fig. 5.27 Number employees per 1000 connections, 1993–2019. *Source* PPWSA

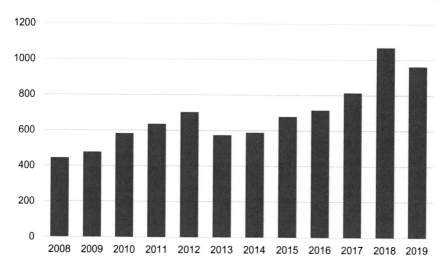

Fig. 5.28 Employee cost per month in USD, 2008–2019. *Source* PPWSA

2.6 in 2012. The employees per 1000 connections in 2009 was 5.9. This statistic was the lowest in 2012 at 2.58. In 2019, PPWSA total employees and employees per 1,000 connections were 1092 and 2.8 respectively.

Total average employee cost per month increased from USD 445/employee in 2008 to USD 963/employee by 2019 (Fig. 5.28).

The starting government salaries in Cambodia are USD 100–150/month and the private sector salaries are USD 300/month. PPWSA starting salary is somewhere in between the government salary and the private sector salary. The starting salary of an engineer in PPWSA was USD 250/month, excluding other benefits, in 2017.

5.3.3 Unit Realisations

PPWSA's average unit realisation in 1997 was KHR 625/m^3 of water sold, and after the 2001 tariff increase, PPWSA's unit realisation went up to KHR 949/m^3. Though there had been no tariff increase between 2001 and 2016, the average realisation of PPWSA went up by 11% as compared to 2001 due to the block tariffs. For example, the average water consumption for domestic customers per month went up from 21.67 m^3 in 2002 to 27.0 m^3 in 2016. Reason for higher consumption per domestic consumption per month is steadily improving lifestyles and increase in average pressure in the distribution mains from 1.0 to 2.0 bar to 2.5 to 3.0 bar. The average household domestic water bill in 2016 was KHR 25,380/month (USD 6.19). The unit realisations for domestic, commercial and government consumers from 1997 to 2019 are shown in Fig. 5.29.

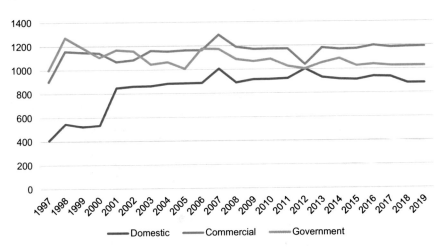

Fig. 5.29 Unit realisations of water sold in KHR/m³, 1997–2019. *Source* PPWSA

A new tariff scheme was implemented on 15 May 2017. The 2001 Tariff and May 2017 Tariff scheme is shown in Table 5.2.

As per the 15 May 2017 tariff schedule, the first block of water consumption between 0–7 m³/month was divided into two blocks: 0–3 m³ and 4–7 m³. In this tariff scheme, the tariff for 0–3 m³ water consumption bracket was reduced from

Table 5.2 PPWSA Tariffs 2001 and 15 May 2017[a]

Customer category	Water used (m³/month)	Old tariff (2001) (KHR/m³)	Proposed tariff (May 2017) (KHR/m³)
Domestics	Up to 03 m³	550	400
	4–7 m³	550	500
	8–15 m³	770	770
	16–50 m³	1010	1010
	Over 50 m³	1270	1270
Government institutions and distributors	Not linked to quantity consumed	1030	1030
Commercial, autonomous state, authorities and retailers	Up to m³–100 m³	950	950
	101–200 m³	1150	1150
	201 m³–500 m³	1350	1350
	Over 500 m³	1450	1450
Landlords (room rental)	Not linked to quantity consumed	1030	1030

Source PPWSA
[a]Later in September 2017, the May 2017 tariff change was cancelled by PPWSA

Table 5.3 Illustration of total bill under old and the May 2017 tariffs schemes (for 18 m³ monthly water consumption)

Bill as per 2001 tariff scheme			Bill as per May 2017 tariff Scheme		
Volume (m³)	Applicable tariff (KHR/m³)	Amount (KHR)	Volume (m³)	Applicable tariff (KHR/m³)	Amount (KHR)
7	550	3850	–	–	–
8	770	6160	–	–	–
3	1010	3030	–	–	–
Total	18	–	13,040	18	1010	18,180

Source PPWSA

KHR 550/m³ to KHR 400/m³ and tariff for 4–7 m³ of water consumption bracket was reduced from KHR 550/m³ to KHR 500/m³.

However, in the tariff scheme that was implemented from 15 May 2017, there was a significant difference in the method of tariff calculation. For example, if a household consumed 18 m³ of water a month, in the earlier tariff scheme of 2001, it would have been charged KHR 550/m³ for the first 7 m³ of water consumed, and then KHR 770/m³ for the next incremental 8 m³, and KHR 1010/m³ for the balance 3 m³ of water. However, under the tariff scheme implemented from May 2017, the household would have to pay KHR 1010/m³ for the entire 18 m³ of water consumption. In other words, the tariff scheme, there was a 39.4% increase in water bill for a domestic household consuming 18 m³ of water. An illustration of water bill for 18 m³ of monthly water consumption by a household under the 2001 tariff scheme, and under the May 2017 tariffs, is shown in Table 5.3.

Many customers complained about the sudden jump in their monthly water bills. Following these complaints, Prime Minister Hun Sen instructed PPWSA to roll back the new tariff structure. Accordingly, PPWSA rolled back the May 2017 tariff scheme and retained the old tariff scheme, with some minor changes to what we call here as revised May 2017 Tariffs. These new changes included the "rented room" segment for whom the water tariff was reduced from KHR 1030/m³ to KHR 700/m³. The revised May 2017 Tariffs also reduced standpipe water tariffs from KHR 1030/m³ to KHR 700/m³.

The "rented room" segment of customers as well as the Standpost segment each consume 5% of the total volume sold by PPWSA. A 32% drop in the tariff, as per the revised May 2017 tariffs, for both these segments would have meant a cumulative negative impact of 3.2% on the revenues of PPWSA.

A new tariff scheme was announced in Sep 2019 with an effective date of implementation being 1 Jan 2020[1] what we call as Jan 2020 tariffs. The progression of the tariff structures from 2017 to 2020 is shown in Table 5.4.

The January 2020 tariff scheme devised new consumption categories in domestic as well commercial segments. In the domestic segment, a category of 0–7 m³ replaced

---

[1] Annexure IV provides the official declaration of the tariffs implemented from 1 Jan 2020.

Table 5.4 Progression of tariff structure from 2017 to 2020

Customer category	Consumption (m³/month)	Tariff before 15–05-2017 (2001 tariffs) KHR/m³	Tariff from May-2017 to Dec-2019 (Revised May 2017 tariffs) KHR/m³	Tariff from 1 Jan 2020 (Jan 2020 Tariffs) KHR/m³
Domestic	0–03 m³	550	400	400
	04–07 m³	550	500	400
	08–15 m³	770	770	720
	16–50 m³	1010	1010	–
	Over 50 m³	1270	1270	–
	16–25 m³	–	–	960
	26–50 m³	–	–	1250
	51–100 m³	–	–	1900
	over 100 m³	–	–	2200
Government institutions and distributors/Administrative	Not linked to quantity consumed	1030	1030	2500
Commercial, autonomous state authorities and retailers	Up to 15 m³	–	–	950
	16–45 m³	–	–	1100
	46–100 m³	–	–	1400
	Up to 100 m³	950	950	
	101–200 m³	1150	1150	1700
	201–500 m³	1350	1350	2100
	Over 500 m³	1450	1450	2400
Landlords (room rentals)	Not linked to quantity consumed	1030	700	

Source PPWSA

the 0–4 m³ category and the tariff for this segment is now KHR/m³ 400. The overall impact of the change in domestic tariffs is that up to 25 m³ monthly consumption the water tariff has been reduced in the range of 5–20%. However above 25 m³ monthly consumption, the domestic customer category tariffs have gone up by 24–73%. In the commercial category also, new consumption categories have been introduced. The range of water tariff increase for the commercial consumption categories has been 16–65%. As per Jan 2020 tariffs, the administrative tariffs have gone up by more than 142%. Jan 2020 tariffs left the water tariffs the "Rented room" and Standpost categories unchanged from the previous levels.

In absence of greater details of the volume sold in each consumption category it is difficult to guess the effective hike in the Jan 2020 tariffs as compared to May 2017 water tariffs. A crude estimate of effective tariff for PPWSA in 2020 as compared to

2017 could be in the range of 13–20% or KHR/m^3 1200–1300. Though the increase in the tariffs is a welcome step, PPWSA will need to continuously increase its water tariffs to KHR/m^3 1600–1900 which we estimate as PPWSA marginal cost of water supply.

5.3.4 Financial Ratios

The water price remained stationary between 2001 and 2016. However, costs of production and distribution of water increased steadily during this period. Thus, the earnings before interest, depreciation and amortisation and tax (EBITDA) margins, as well as earnings before interest and tax (EBIT) margins, had declining trends from 2008 to 2016. The margins recovered in 2017 and 2018 but have again declined in 2019. The trend of EBIT and EBITDA margins is shown in Fig. 5.30.

Though the profitability margins have gone down, the asset turnover ratios of PPWSA remained relatively stable till 2018. However, the asset turnover ratios dropped in 2019 (Fig. 5.31).

The ratios of profitability on total capital employed (EBIT/total assets) have been fluctuating at a narrow range, between 3.8 and 4.9% between 2008 and 2016. The same went up but again declined to their lowest levels in 2019 in the time period 2008–2019 (Fig. 5.32).

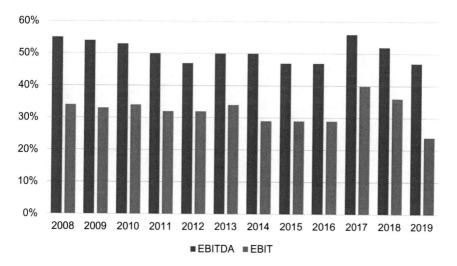

Fig. 5.30 EBITDA and EBIT margins, 2008–2019. *Source* PPWSA

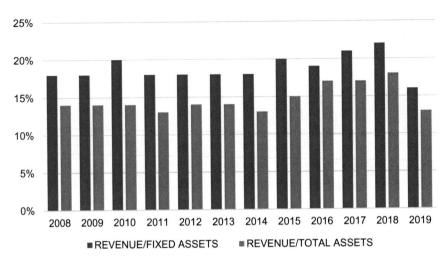

Fig. 5.31 Asset turnover ratios, 2008–2019. *Source* PPWSA

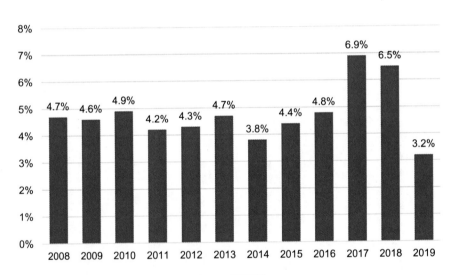

Fig. 5.32 EBIT/total assets, 2008–2019. *Source* PPWSA

Theoretically, the utility businesses can increase debt to equity ratio to 70:30%. The debt to equity ratio of PPWSA in 2016 was 27:73%. This may indicate that PPWSA has capacity to borrow more and put this debt on its balance sheet. However, given that the profitability of the business can come under stress if tariff rates are not revised at regular intervals, it may not be prudent for PPWSA to increase its financial leverage.

5.3.5 *Social Fund by PPWSA*

The social fund was established in May 2008, following the approval by the Ministry of Economy and Finance as its financial guardian ministry. This fund is derived from retention of 5% of the annual profits. In terms of accounting this is recorded as expenses in the year that these expenses are incurred. This fund has been administered by the Social Fund Committee appointed by the PPWSA Director General.

The objectives of the social fund are to ensure good relations exist between PPWSA and the society through subsidies to charitable activities, focusing on poverty reduction, and without any political or religious discrimination.

Currently, this fund has been classified into four categories of expenses: construction of social infrastructure (50% of total social fund), expenditure in public interests such as wells, potable water systems, bridges, roads, rest rooms, schools, and so on. Some 30% for direct/indirect support to the people in need, such as victims of natural disasters, subsidies to poor students, and 10% for help to low-income customers, water connections and other expenses. The balance of 10% of the social fund is used for contingencies.

The average costs of water sold per m^3 in USD from 1996 to 2019 are shown in Fig. 5.33.

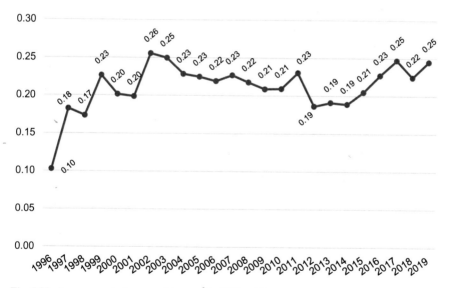

Fig. 5.33 Average cost of water sold per m^3 in USD, 1996–2019. *Source* PPWSA

5.3.6   *Financial Outlook*

As per PPWSA Master Plan 2016–2030, the marginal cost of water from new water treatment plants is estimated to be KHR 1900/m^3. This is almost twice the current cost of production of water for PPWSA. Accordingly, the water utility is in dire need to increase its profitability. The institution has been able to drive all the financial benefits in the past by steadily improving its operational efficiencies. It has little scope left for any additional significant financial savings by improving its internal operational matrices. Furthermore, its non-revenue water has been creeping up steadily since 2012. This is further reducing the utility's total revenue. Arresting this increase in NRW is likely to need additional financial resources for both capital investments and additional operating costs. All these developments mean that after nearly two decades of providing water at the same tariffs, PPWSA has revised tariffs in 2017 and most recently from 1 Jan 2020. The latest tariffs will ameliorate some of the financial stress. PPWSA will eventually need to raise the water tariffs to cover its marginal cost of water supply.

5.4   Institutional Domain

This section provides an objective discussion of the various government agencies that are responsible for the Cambodian water sector in general, and PPWSA in particular. It also presents an overview of the important plans and policies of the water sector in Cambodia.

5.4.1   *Water Agencies in Cambodia*

Under Article 3, Chapter 1, of the Law of Water Management of the Kingdom of Cambodia, all water and water resources of the country are owned by the state. The term "water" in the context of this law includes surface water, groundwater and atmospheric water. Further, under Article 11, Chapter III, of the same law, stipulates that every person has the right to use water resources for their vital needs including drinking, washing, bathing and other domestic purposes like water for animal husbandry, fishing and irrigation of domestic gardens and orchards. However, this has to be accomplished in a manner that will not affect the legal rights of others. The amount used should not exceed what is necessary. The aforementioned uses are not subject to any licensing.

Like in all other countries, there are multiple government agencies that are involved in managing the Cambodian water sector. For example, The Ministry of Rural Development (MRD) is responsible for supplying water to the rural communities. The diversion, abstraction and use of water resources for purposes other than those mentioned in Article 11 of the Law of Water Management, and the construction of the waterworks related thereto, are subject to a license or permit that has to be issued by Ministry of Water Resources and Meteorology (MOWRAM). Prior to the granting of a use license to any person, this Ministry has to consult with other appropriate agencies and the local authorities that are concerned with the water utilisation and the construction of waterworks proposed by the applicant. Water use licensees may transfer their water use rights totally or partially to another user, after securing the prior approval of MOWRAM. However, MOWRAM has the right to cancel/modify the existing abstraction licenses for valid reasons. This Ministry is also responsible for administration and management of groundwater in the country.

The Ministry of Public Works and Transportation (MPWT) is responsible for wastewater as well as stormwater drainage. The Ministry of Environment (MoE) is responsible for preventing pollution of the water bodies. The Ministry of Industry and Handicrafts (MIH) is responsible for drinking water supply in urban areas. This Ministry is further responsible for establishing water quality standards. It is also responsible for granting licenses to private water operators in the urban areas. Thus far it has given licenses to 226 private Cambodian operators. Licensees have to pay an annual license fee, 51% of which goes to the Ministry of Economy and Finance and 49% goes to MIH. The Ministry of Economy and Finance (MEF) is responsible for all the foreign loan negotiations that may be used by the water utilities of Cambodia like PPWSA. Various ministries involved administering the water sector are shown in Fig. 5.34.

MIH is the reporting ministry for PPWSA and 12 other public water utilities of Cambodia. Table 5.5 provides further details of government agencies that are responsible for key functions and their relevance and implications for PPWSA.

5.4.2 Ministry of Industry and Handicrafts, Now Ministry of Industry Science Technology and Innovation

The Ministry of Industry and Handicrafts (MIH) has been the tutelary ministry of PPWSA since 2014. Prior to this date, the Ministry of Mines and Energy (MIME) was the tutelary ministry from 2004.

MIH regulates the public and private water utilities of Cambodia. Currently, MIH regulates 13 public water utilities, including PPWSA. It regulates and licenses private water utility operators. There are 500 private water utility operators and 226[2] of the 500 are licensed private operators. Furthermore, as noted earlier, the Director General of the General Department of the Ministry of Industry and Handicraft is the chairman of the board of directors of PPWSA.

Prior to 2006, there was only one department called the Department of Potable Water in the Ministry of Mines and Energy that looked after all the Cambodian public and private water operators. In 2006, the Department of Potable Water had

---

[2]MIH Interviews (September and December 2017).

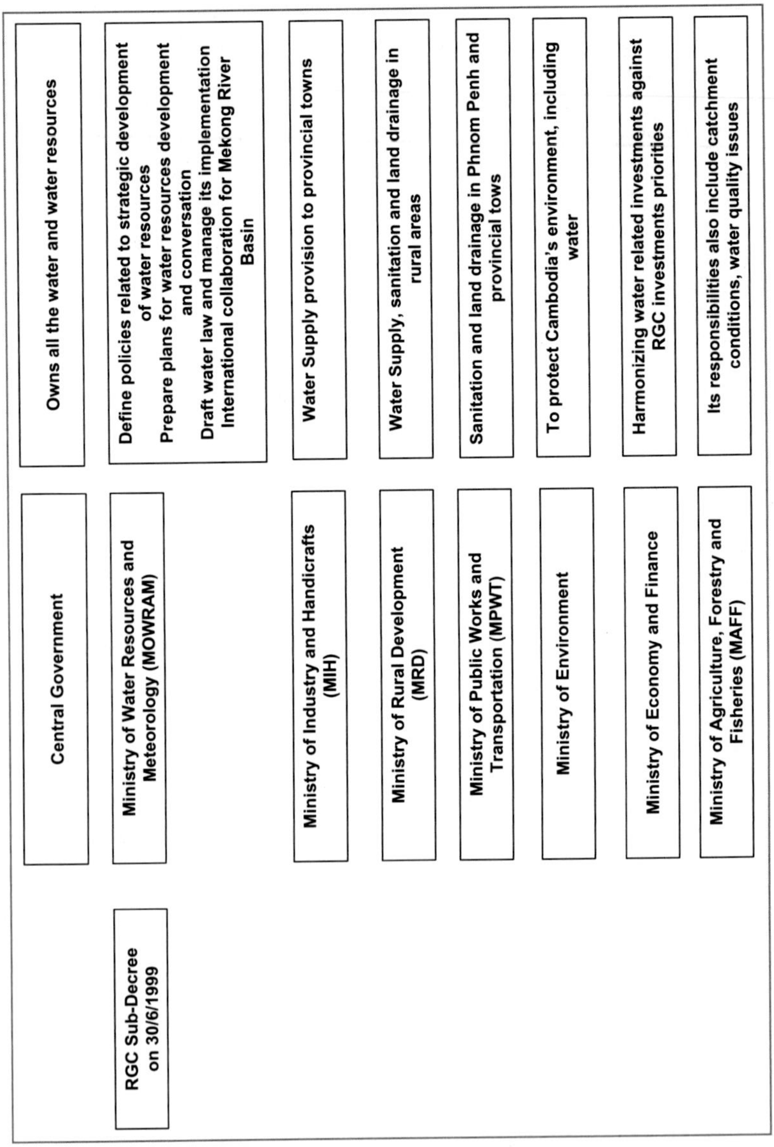

Fig. 5.34 Ministries administrating Cambodian water sector

RGC Sub-Decree on 30/6/1999

Central Government — Owns all the water and water resources

Ministry of Water Resources and Meteorology (MOWRAM)
- Define policies related to strategic development of water resources
- Prepare plans for water resources development and conversation
- Draft water law and manage its implementation
- International collaboration for Mekong River Basin

Ministry of Industry and Handicrafts (MIH) — Water Supply provision to provincial towns

Ministry of Rural Development (MRD) — Water Supply, sanitation and land drainage in rural areas

Ministry of Public Works and Transportation (MPWT) — Sanitation and land drainage in Phnom Penh and provincial tows

Ministry of Environment — To protect Cambodia's environment, including water

Ministry of Economy and Finance — Harmonizing water related investments against RGC investments priorities

Ministry of Agriculture, Forestry and Fisheries (MAFF) — Its responsibilities also include catchment conditions, water quality issues

Table 5.5 Government Agencies and their functions in water sector

Function	Government Agency responsible for the function	Relevance for PPWSA
Surface water abstraction rights	MOWRAM	PPWSA does not require approval from anyone for volume of surface water abstracted from Mekong, Tonlé Sap or Bassac rivers
Drinking water quality standards	MIH	MIH has prescribed 47 parameters for ensuring drinking water quality standards. PPWSA must monitor them
Economic regulation	MIH	In the final analysis, the Prime Minister of Cambodia decides on the water tariffs that PPWSA can use
Consumer interests		PPWSA is not answerable to any government agency for consumer interests
Foreign loans and grants	MEF	All the foreign loans of PPWSA are routed through MEF since overall national debt has a ceiling

Source PPWSA

only 26 staff members. In July 2016, a new General Department of Potable Water Supply was formed within the MIH. This has five internal divisions: Administration, Planning and data management, Potable water supply policy, Technical and project management and Regulation. Currently there are about 60 staff members working in this Department. The Regulation Department has 12 staff members and 8 of them are engineers.

From 2014, MIH has been entrusted with establishing water tariffs for all public and private utilities that it regulates. It is further responsible for establishing drinking water quality standards. It also has a mechanism for registering customer complaints. However, this is still in its nascent stage.

MIH at present is seriously understaffed to carry out all the tasks they are entrusted with. In addition, it urgently needs capacity enhancement. To the credit of many donors, they are assisting with building institutional capacities. However, much more remains to be done.

5.4.3 Water Sector Plans and Policies

National policy on water supply and sanitation (PPWSA 2017b).

The national policy on water supply and sanitation was prepared jointly by the Ministry of Industry, Mines and Energy and the Ministry of Rural Development. The policy received legislative approval in February 2003. The policy has three components dealing with urban water supply, urban sanitation, and rural water supply and sanitation. Many donor agencies have conducted water sector policy assessments for Cambodia. In one of such assessments, ADB noted that there was a lack of definition for minimum level of services for both water supply and sanitation. ADB further noted lack of clarity for demarcations between urban and rural areas. There was also a lack of clarity between the roles and the responsibilities of MIME and MRD. As per the World Bank water sector review of Cambodia (2012), many provisions and laws of the February 2003 policy were yet to be implemented.

Law on water supply, (Draft, 2004) (PPWSA 2017b).

A law governing the provision for water and sanitation services was drafted in 2004, but it was never formally adopted. The water sector in Cambodia currently does not have a comprehensive legal framework for an efficient and functional institutional arrangement which could assure rational and sustainable planning and management. For example, there is no independent regulator for its municipal water sector.

Law on water resources management (2007) (PPWSA 2017b).

The Law on Water Resources Management (2007) is under the purview of the Ministry of Water Resources and Meteorology. This law stipulated that all the water and water resources in the country belong to the State. Under this law, all commercial uses of water need licenses or permits. The construction of structures over the bed, banks and shores of natural water bodies, including bridges and buildings requires approval by MOWRAM.

5.4.4 Institutional Domain Outlooks

Cambodia clearly needs updated laws on the water sector since the Law on Water Supply (2004) has still not been adopted. As noted earlier, the institutional capacity of MIH needs to be significantly enhanced. For example, MIH does not even have a lawyer in its division which is looking after all their regulations of its water sector. MIH needs to hire more experienced and knowledgeable staff and concurrently build the capacity of its existing staff. The successes of the overall water sector, and also of PPWSA, will largely depend on the capacities of the regulator to appreciate the underlying financial and operational demands of the water utilities. There is a clear need to have increased institutional focus on quality of water as much as on the quantity of water that is available to the households.

References

Liemberger R (2002) Do you know how misleading the use of wrong performance indicators can be? Paper presented at IWA Managing Leakage Conference, Cyprus, November 2002. https://www.geocities.ws/kikory2004/7_Liemberger.pdf

PPWSA (Phnom Penh Water Supply Authority) (2017a) Annual report 2017. Phnom Penh Water Supply Authority, Phnom Penh

PPWSA (Phnom Penh Water Supply Authority) (2017b) Phnom Penh water supply authority third master plan, period 2016–2030, vol. 1 (Master plan report). Phnom Penh Water Supply Authority, Phnom Penh

Chapter 6
Gap Analysis of Four Domains

This section analyses gaps in the physical, operational, financial and institutional domains of PPWSA under the status quo situation. The term "gap" in the current context, stands for what is lacking in each of the domains from the perspective of going from "what it is at present basis" to "what it should be at basis." What it should be at basis is guided by the aspirational goals of the water utility.

In this context, aspirational goals for PPWSA for 2030 are considered as per its third master plan. The overall gap analysis provides a degree of gaps in the physical, operational, financial and institutional domains into three categories in decreasing order of attention required: "Need serious attention"; "Need corrective attention" and "Need attention". The gap analysis summary for PPWSA is shown in Table 6.1.

The institutional domain of PPWSA needs serious attention. MIH is currently responsible for a range of responsibilities, including consumer protection and economic regulation. However, with limited manpower it cannot give adequate attention to all the tasks it is responsible for. There is an urgent need to increase the number of staff and their ability to look after functions such as water quality and tariff setting. MIH needs to build capacity in estimating the appropriate marginal costs of supply of water for water utilities like PPWSA. For example, MIH should be able to calculate about the cost of marginal water under different sets of scenarios, like where only one new water treatment plant capacity is required; or along with a water treatment plant new transmission line also is needed; or along with water treatment plant capacity both new transmission and distribution lines are required. These types of capacities are mostly missing at MIH currently.

The wastewater and drainage in Cambodia are regulated by the Ministry of Public Works and Transportation and drinking water by MIH. Currently, wastewater treatment does not exist in Phnom Penh, or any other urban centres of Cambodia. The water consumption of Phnom Penh is estimated to double from the current levels to 1000 MLD by 2030. This means that 750 MLD of wastewater could be potentially needed to be treated by 2030. Continued discharge of large volumes of untreated wastewater to the environment will have serious implications for qualities of surface and groundwater, as well as on human and ecosystems health. Thus, Cambodia needs

© The Author(s), under exclusive license to Springer Nature Singapore Pte Ltd. 2021
A. K. Biswas et al., *Phnom Penh Water Story*, Water Resources Development and Management, https://doi.org/10.1007/978-981-33-4065-7_6

Table 6.1 Gap analysis of four domains

Domain	Degree of gap
Physical	Need attention
Operational	Need corrective action
Financial	Need serious attention
Institutional	Need serious attention

to consider establishing a single regulatory body, on a priority basis, for drinking water, storm water and wastewater disposal. Such a new institution is likely to have a more holistic and realistic view of the entire municipal water services in terms of environment protection, tariff setting and financial self-sufficiency.

The financial domain of PPWSA needs immediate attention. Currently PPWSA is still profitable. However, its marginal cost of water is expected to be higher than the average unit realisation within the next few years. The marginal cost of the new proposed capacity of PPWSA is expected to be KHR 1900/m^3 and the current unit realisation is nowhere close to that. The proposed new water treatment plant capacity of PPWSA is planned to come online by 2023. PPWSA needs institutional as well as strong political support to increase tariffs in line with its steadily increasing costs to deliver water and good services to its consumers. The tariffs need to cover the marginal cost of water supply.

The operational domain of PPWSA has been praiseworthy during the last decade with NRW consistently less than 8%, and high employee productivity of less than 3.3 employees per 1000 connections. This productivity is especially noteworthy since PPWSA performs nearly all tasks by itself, with no outsourcing of work to the private sector. However, the operational domain of PPWSA needs corrective action since its NRW is slowly increasing. The utility also faces additional difficult challenges since some of its key management team are retiring within the next 3–5 years. The institution has to plan for leadership renewal.

The physical domain of PPWSA is in a good position with ample quantity of water available from the Mekong river. However, this domain still needs attention as the quality of raw water for its Phum Prek and Chamkar Mon can deteriorate further if the wastewater of Phnom Penh continues to be discharged.

Chapter 7
Views of the Customers

Interviews were carried out with selected residents of Phnom Penh about their views and experiences with the providers of different urban services. The sample size of twenty was small and was not randomly selected. However, the interviews provide a good insight into residents' views on the qualities of different urban service they are currently receiving. In this survey, the residents of Phnom Penh were asked about their opinions on the qualities of municipal services they are receiving in the areas of water supply, sewerage, stormwater drainage, electricity supply, solid wastes, and cooking gas supply. Amongst other views, the residents were asked about what changes they would like to see in the water supply and sewerage services and what would be the single most government policy that they would like to see changed. A sample of questionnaire that was used in the survey is given in Annex III.

Amongst all the urban services, the residents were most satisfied with the drinking water supply. All the residents replied that they were receiving 24-h water supply. However, nearly a quarter of the respondents complained about inadequate pressure during the peak hours in first floors and above. Though residents had some idea about how much their water bills in KHR, very few of them knew how much volumes they were consuming every month. Most of the people had no complaints about the water supply, though there were two responders who complained that brown water was coming out from their taps. One respondent also thought the piped water supply had odour. Another respondent complained of diarrhoea. To be on the safe side, this household was drinking water from a 20-L bottled water container. Another respondent said that she only drank bottled water since she did not trust the quality of the tap water she was receiving, especially for drinking.

All the residents felt that the water service has improved in Phnom Penh during the past ten years. One respondent noted that earlier whenever there was electricity cut, there was a stoppage in water supply. This was no longer the case. Another respondent noted that a few years ago there was a problem with the quantity and pressure of water that his household was receiving. However, this has changed. They are now receiving adequate quantities of water with sufficient pressure.

A. K. Biswas et al., *Phnom Penh Water Story*, Water Resources Development and Management, https://doi.org/10.1007/978-981-33-4065-7_7

The suggestions made by the residents for better water supply services included the following. The Government should have well defined policies on health, and also make a determined effort to lower corruption. The Government should ensure water is a high priority item in its agenda. Water supply should be available in the far-off areas of the city with sufficient pressure. Maintenance works should be fast and effective. The utility should adopt more technology. PPWSA management should focus on the quality of water as well as on the quantity. There should be increasing focus on the reuse of water. The utility should give higher emphasis on environmental protection. Only one respondent suggested that the water price should be reduced.

On sewerage services, there was lack of knowledge amongst the respondent as to which agency was responsible for managing these services. Most of them were not aware that 10% of their water bill was for sewerage services. They also did not know what was happening to the wastewater discharged from their households. One common complaint was that the pipelines carrying wastewater were getting clogged.

In terms of suggestions to improve wastewater services in Phnom Penh, respondents felt that the city should have good urban planning and management, and regular maintenance and clearance of pipes to avoid clogging and bad smell. Solid wastes collection in the city should be improved, and the residents should be informed if these wastes are creating problems for wastewater collection networks. The authorities should prevent landfilling the areas where the wastewater is being discharged. The capacities of the pipes carrying wastewater should be increased. The Government should listen to the criticisms from the citizens and lower corruption and rent-seeking practices.

On drainage, some respondents said they thought that drainage infrastructure has improved recently. However, many of the respondents said that during the rainy season they encounter flooding to depths which could vary from ankle to the shin levels. Such flooding lasts from 30 min to one day. One respondent complained that whenever the manholes are opened during the rainy seasons, there are backflows of wastewater and stormwater on the road. This, in turn, caused flooding with very poor quality of water which is a health and environmental hazard.

The respondents' responses were not necessarily sanguine on solid wastes management. They felt that solid wastes were collected by the operator, sometimes only once a week. Some others felt that the pricing structure of the solid wastes' collector was not transparent. Most of the respondents were aware that they were paying solid waste charges along with their electricity bills.

On electricity, the respondents felt services have improved during the past ten years. However, there are still blackouts and inefficiencies. The most common complaint was the high electricity price. The topmost wishes of the respondents were that electricity prices should be reduced, and supply should become more consistent. They felt that they were unable to use all the electrical appliances they would have liked as there was inadequate power supply, and the electricity prices are far too high.

The respondents were happier with the domestic water supply services compared to electricity. They especially liked the attitudes of PPWSA staff who came to bill them. In this connection, it is appropriate to note an anecdotal comment by one of the respondents on the overall performance of PPWSA. He asked: *Is the water supply company of Phnom Penh a private company or public company?* On being told that the utility was a public company, his response was: *How can it be such a good company since it is a public one?*

Chapter 8
Key Challenges Facing PPWSA

8.1 Human Resource Planning and Implementation in PPWSA

An important reason for the success of any organisation is the quality and the commitment of the people who work for it. The current top management team at PPWSA is a very experienced one. Most of the Deputy Director Generals have been with the institution for more than two decades. Some of the Deputy Director Generals had studied in countries outside Cambodia. However, most of the current senior management staff will retire within the next five years. There is thus an urgent need to build the next generation of leaders so that they can seamlessly move into senior management positions as and when the current managers retire. PPWSA can also consider hiring retiring senior management staff as advisors on a part-time basis.

PPWSA also needs to prepare a comprehensive long-term, future-oriented, recruiting plan for new staff, which should be prepared in consultations with all the departments as well as its supporters and advisors. The recruitment plan should consider the likely competition in the Cambodian job market because of significant competition for new and experienced staff from both public and private sectors. Rapid economic growth in the country will make this competition, to have and retain good staff, progressively more and more intense.

The human resource planning also faces challenges of somewhat lower quality of school and college education that currently exist in Cambodia. Unlike the earlier generation of management leaders who went and studied engineering outside Cambodia during the 1980s and the 1990s, and then returned to work in Cambodia, there does not appear to be much interest amongst the current younger generation to go abroad for engineering degrees. This may be because jobs are plentiful for the current generation of graduates compared to earlier generations. Even when the younger generation goes abroad to study, their first choice often tends to be to find employment outside Cambodia, and thus not return to the country.

The current educational system in schools has often suffered from widespread cheating and corruption. Admittedly, the school education system has been

© The Author(s), under exclusive license to Springer Nature Singapore Pte Ltd. 2021
A. K. Biswas et al., *Phnom Penh Water Story*, Water Resources Development and Management, https://doi.org/10.1007/978-981-33-4065-7_8

improving, especially during the last five years. A crackdown on cheating started in 2014. This resulted in the 12th grade pass rate improving from 27.5% in 2014 to 63.84% in 2017. In absolute numbers, only 63,666 people passed the 12th grade examination all over Cambodia in 2017 (Koemsoeun and Handley 2017).

As per the official government figures for 2016–2017, 47.9% of schools in Phnom did not have access to clean water and 26.1% of schools did not have toilets in their premises. There is thus an urgent need to improve the provision of water and sanitation services in educational institutes even in the capital city. The situation outside the capital is generally worse.

Overall PPWSA faces a major challenge in replacing the current experienced and competent management team with equally good candidates within only less than five years. The overall quality and infrastructure of Cambodia require steady improvements. Thus, not only replacing retiring employees is likely to be a problem but also hiring costs of new employees are most certainly going up as there will be serious competition to hire competent employees from both public and private sectors from a steadily expanding economy, both in Cambodia and also in the region.

8.2   Phnom Penh Urban Master Plan

A city should have three main goals. These relate to fulfilling its economic, social and environmental goals. The economic goals are to have a robust, vibrant and sustainable economy. The social goals are to provide good quality of living and sense of well-being to all. The environmental goals are to protect and manage the environment the best way possible. A city can fulfil all these three important goals if it has a detailed long-term Urban Master Plan which could be updated every five years depending on changing data and assumptions during the ensuing five years. Furthermore, a plan is not enough: determined and sustained efforts should be made to implement it. This Urban Master Plan should detail the present and the future zoning of residential, commercial and industrial areas, and provide details of the area dedicated to transportation, waterways, green spaces, etc. A long-term planning by a water utility is dependent on having a sound urban master plan for that city.

PPWSA faces major challenges with reference to Urban Master Plans of Phnom Penh. From 2002 to 2007, the Bureau of Urban Affairs of the Municipality of Phnom Penh, prepared an Urban Master Plan, with 2020 as the planning horizon (PPWSA 2017). This document was never adopted, let alone implemented. In 2015, it was announced that the Urban Master Plan had been extended until 2035. The final document "Phnom Penh Land Use Plan for 2035" was adopted by the Council of Ministers on 11 December 2015 (PPWSA 2017).

However, most unfortunately, PPWSA was neither consulted nor requested to make comments during the preparation of the Urban Master Plan. Accordingly, its appropriateness and relevance in terms of water planning was not considered (PPWSA 2017). This has meant PPWSA has been seriously handicapped to formulate its own long-term plan to ensure water security of Phnom Penh for the 2016–2030 period.

In addition, PPWSA does not have reliable data on the current population of Phnom Penh, and on the number of people living in each household. With rapid urbanisation, Phnom Penh is growing in all directions. The city does not have any specific land use plans. Nor does PPWSA have details and relevant information on which parts of Phnom Penh are likely to grow the most during the next decade. Without such essential information, making a long-term useful water plan is very difficult under the best of circumstances.

In retrospect, the hydraulic design capacity of the water pipeline network of PPWSA was compromised in 2012. This is because twenty districts of the Kandal province were merged with Phnom Penh. The action, in a political sense, is understandable. Amalgamation of these twenty districts to the area served by PPWSA made good political sense since all the households in these new districts did not have good and reliable water supply. Amalgamation of these districts to Phnom Penh allowed all their households to have prompt access to clean water. The original design capacity of the water network was to add 50 km of distribution pipeline and 5000 connections every year. However, due to the expansion of the city limits and ensuing strong political pressure, the distribution network had to be expanded by 200 km of additional distribution lines and further additions of 20,000 connections every year. This has compromised the quality of the service, especially in terms of both the pressure and the quantity of water that is being received by the most households in these expanded areas.

This is an important development which should not have occurred had the responsible authorities consulted with PPWSA for the formulation of the Urban Master Plan of Phnom Penh. Even then, it should be noted that the Urban Master Plan of Phnom Penh does not have any provision for land for the future water treatment plants that have to be built for Phnom Penh. PPWSA has been able to purchase a 15-ha parcel of land for its new proposed expansion, at a cost of USD 25 million. This has come from its own internal budget. The land for water abstraction for the new expansion has also been procured at a further cost of USD one million. There has been no concession, or no free land, that was made available to the utility to build its future water treatment plant.

Total lack of consultation with PPWSA during the formulation of the Urban Master Plan will continue to remain a serious handicap to PPWSA for its future own long-term planning. In order to prepare a realistic and feasible Urban Master Plan PPWSA should have been extensively consulted during the Plan preparation, since water supply is one of the essential infrastructure requirements for the city. The issue is likely to haunt future urban planning of Phnom Penh. It is now a moot point whether the lack of PPWSA involvement of any type during the Urban Master Plan formulation was an accident, or by design, perhaps because of inter-institutional jealousy or rivalry. If it is the later, then there is likely to be more problems for PPWSA in the coming years. In either case, the current situation is not desirable.

8.3   Evolving Institutional Domain

The Ministry of Industry and Handicrafts (MIH) is responsible for the economic regulation of the public as well as the private water utilities. There are close to 500 private water operators in Cambodia, of which 226 are currently licensed by MIH. In addition, MIH regulates 13 public water utilities. It is further responsible for economic regulation, establishing water quality standards and handling consumer complaints. This Ministry has a wide range of responsibilities and yet its regulation department, in 2017, had only twelve employees. It undoubtedly needs more experienced staff if it is to manage successfully all the responsibilities it has been assigned. It also needs to employ people with diverse disciplinary backgrounds and then forge them into one team. Of the 12-member staff working in the regulation department of the General Department of Potable Water, in the Ministry, eight employees are engineers. The regulation department somewhat surprisingly, does not even have a single lawyer.

PPWSA has projected a total capital cost of USD 414 million from 2018 to 2025 to build new water treatment plants and new transmission and distribution networks (PPWSA 2017). The nature and scale of the future projects of PPWSA are going to become increasingly more complex and costlier than ever before. The Ministry needs to develop adequate expertise to manage such complex and difficult projects.

The marginal cost of water supply from new water treatment plants is expected to be KHR 1600–1900/m^3 (PPWSA 2017), as against estimated current unit sales realisation of KHR 1200/m^3. MIH needs to upgrade its skills to calculate the marginal cost of water not only for Phnom Penh but also for all other urban water utilities in Cambodia. It also needs to build capacity to estimate the marginal cost of water under different scenarios related to raw water availability, number of customers, volume of water demanded by end customers and related capital investments.

At present, the Ministry of Water Resources and Meteorology is responsible for managing water resources. However, it does not regulate the quantity of water abstracted by the various water utilities. This Ministry also does not communicate the quality of raw water to the water utilities such as PPWSA. There is an arguable case for the Ministry to regulate the quantity of raw water abstracted by different water utilities. It should also be responsible for informing the quality of raw water regularly to the utilities, and also to the general public.

PPWSA has mostly maximised all its internal efficiencies like reduction in NRW and number of employees per 1000 connections which has plateaued 3.00 since 2010 (Fig. 8.1). In future, PPWSA's direction will largely depend on independent economic regulation by MIH, which should provide reasonable financial returns to PPWSA so that its profitability can be maintained, and it has the financial ability to fund its future capital expenditure needs. This is because the average household bill has plateaued around USD 11 since 2015 (Fig. 8.2). If MIH is unable to provide adequate financial returns to PPWSA in the near future, it may become financially challenged. This will not only jeopardise PPWSA's existing capability of providing

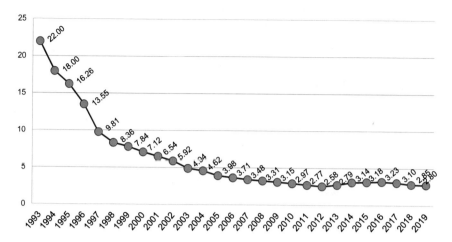

Fig. 8.1 Number of employees per 1,000 connections, 1993–2019. *Source* PPWSA

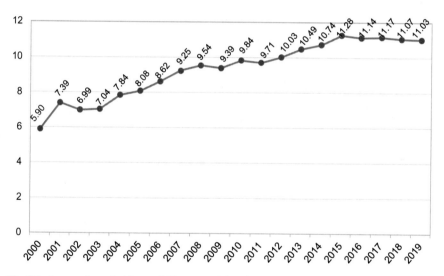

Fig. 8.2 Average household water bill per month in USD, 2000–2019. *Source* PPWSA

good and reliable water services to the inhabitants of Phnom Penh, but also will seriously endanger its medium to long-term financial sustainability.

8.4 Need for Continuous Water Tariff Increase

There are three possible sources of finances for a water utility—funding from the government (funded by taxes), transfers/grants from donor agencies or the water tariffs. At present, PPWSA receives no direct funding from either the local or the national government. It is highly unlikely that PPWSA will receive financial subsidies from the local and national government sources in the coming years. The overall budgets of national and local governments are already severely stretched and are now facing serious funding constraints to provide for other social infrastructure services, not only in Phnom Penh but also for other parts of Cambodia. The problems have further been exacerbated due to unexpected heavy expenditures due to COVID-19 incidences.

PPWSA received a meaningful grant amount in the 1990s from the international donor agencies. However, with the improvement of economic status of Cambodia over the years and by virtue of the fact that PPWSA is now a listed company, it is not possible for PPWSA to secure funding as grants. This leaves tariffs as the only recourse to raise its revenues and securing its financial sustainability.

PPWSA needs to have tariff increases that can beat the inflation. According to the World Bank data, Cambodia's annual inflation rate has varied from a low of 0.60 in 2001 to a high of 24.99 during 2008 (Fig. 8.3). During 2010–2018, annual inflation has fluctuated around 3% rate. This has meant that for PPWSA, the cost of doing business has steadily increased over the past 19 years but its water tariff had remained the same between 2001 and 2017 period.

PPWSA has indeed raised tariffs from 1 Jan 2020, which is a welcome step. However, the tariffs need to reach the marginal cost of water supply and hence the water tariffs need to increase at the regulator intervals from current estimated levels of KHR/m^3 1200 to KHR/m^3 1600–1900 in the next 3–5 years.

The tariff hike by PPWSA should address the problem that it faces in terms of increasing per capita daily water consumption in Phnom Penh (Fig. 8.4).

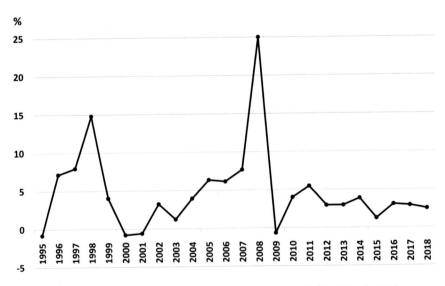

Fig. 8.3 Annual consumer prices inflation in Cambodia, 1995–2018. World Bank (2019)

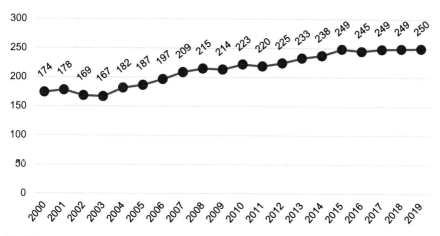

Fig. 8.4 Per capita daily water consumption in litres in Phnom Penh, 2000–2019. *Source* PPWSA

If one considers another Asian city, Singapore, its daily per capita water consumption has steadily declined in recent decades. This has been made possible by encouraging Singapore to understand and appreciate the importance of water conservation at household levels, and recently during 2017–2018 through pricing signals. Singapore's water pricing had remained the same since 2000. It was finally increased by 15% in 2017, and then by another 15% by 2018. The steady decline in daily per capita consumption is shown in Fig. 8.5.

Tariff hike implemented with effect from 1 Jan 2020 is a step in the right direction. However, PPWSA will need to continue hiking its tariffs to reach a weighted average realisation that is equivalent to the marginal cost of water supply. This will help PPWSA not only improve its financial stability but also reduce the rising per capita water consumption in Phnom Penh.

Another reason for PPWSA to continuously increase its tariffs is that it has probably milked most of the possible operational efficiency that it could. PPWSA has managed to remain profitable during this period by improving steadily its operational and management efficiencies. However, much of the easy efficiency gains are now over. For example, its bill collection ratio, that is quantity billed/water produced, was 28% in 1993. It has been consistently above 91% from 2006 (Fig. 8.6). Similarly, its bill collection ratios, that is number of bills collected/total number of bills and amount collected/amount billed, have been consistently around 97–99% since about 2003. This is shown in Fig. 8.7.

Furthermore, the number of functional meters has been consistently around 97–99% since 2003 (Fig. 8.8).

Accordingly, the utility now has very limited scope for improving its operational efficiencies further in the coming years. Thus, for PPWSA to remain profitable and continue to increase its water production and still provide a good service to its consumers, it has no alternative but to increase its water tariffs in the coming years.

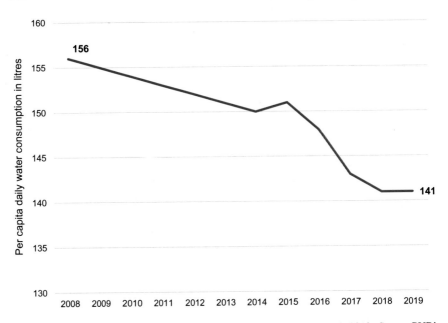

Fig. 8.5 Per capita daily water consumption in litres in Singapore, 2008–2019. *Source* PUB's Annual Reports; Key Environmental Statistics published by the Ministry of the Environment and Water Resources (MEWR)

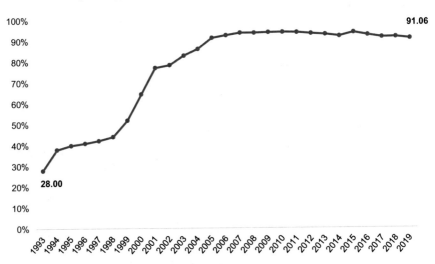

Fig. 8.6 Bill collection ratio (quantity billed/water produced), 1993–2019. *Source* PPWSA

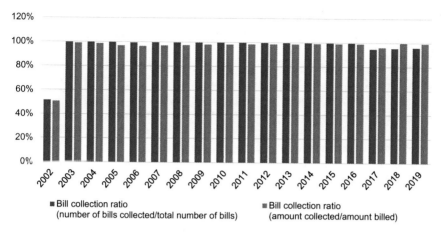

Fig. 8.7 Bill collection ratios (bills collected/total number of bills and amount collected/amount billed), 2002–2019. *Source* PPWSA

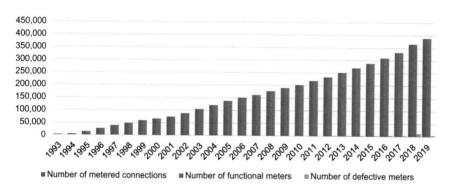

Fig. 8.8 Number of functional meters, 1993–2019. *Source* PPWSA

8.5 Lack of Centralised Sewerage System

Currently there is no centralised sewerage system in Phnom Penh. The sewage and stormwater eventually find their way to the local lakes (*boeng* in Khmer means lake). The areas of these lakes and swamps, where untreated wastewater eventually drains to, are being reduced very fast due to land reclamation to account for increasing urbanisation and industrialisation (JICA 2016). As the lake and the swamp areas are steadily being reduced, and quantities of wastewater discharged increases with time, the city is witnessing higher frequencies of flooding during the monsoon rains, and also durations of floods containing wastewater are increasing. If these trends continue, stormwater flooding in the city will progressively become worse. Changes in rainfall patterns due to climate change in the coming years are likely to further worsen the situation.

The maximum daily water consumption is estimated to increase from 560 MLD in 2016 to 1000 MLD by 2030. If no proper wastewater management system is introduced in Phnom Penh, including collection, proper treatment and safe disposal of domestic, commercial and industrial wastewaters, around 700–800 MLD[1] untreated wastewater will be discharged within and near the city by 2030. This will severely contaminate the water bodies and groundwater in and around the city, with likely adverse impacts on the health of the people and the ecosystems.

JICA has prepared a framework for sewage management plan for Phnom Penh Municipal Administration that is expected to be implemented by 2035. The sewage treatment plan is planned to be executed in stages. Total project cost is estimated at USD 1025 million, with the first phase covering the area around the Cheung Ek Lake Area of Phnom Penh. This first phase will include a sewage treatment plant (STP) of 282 MLD capacity. The first phase is expected to cover sewage generated from an area of 4,701 ha, which currently has a population of 1.09 million. The construction cost of the proposed STP covering the Cheung Ek Lake Area, during the first phase, is estimated to cost USD 459 million. In addition, this will incur an annual operation and maintenance costs of USD 14.89 million. It is further estimated that a wastewater management charge of 75% of water tariff will be necessary to cover only the operation and maintenance costs of the new wastewater management system. This does *not* include the investment costs.

A pre-feasibility pilot study is being conducted to construct a sewage treatment plant of 5 MLD capacity with a sewer pipe network length of 1.3 km. The cost of this pilot project is estimated at USD 24.05 million, with an annual operation and maintenance costs of USD 0.41 million.

In addition to the wastewater treatment plan for 2035, JICA has also prepared a plan for proper stormwater management by 2035. Its total cost is estimated at USD 662 million, with an annual operation and maintenance cost of USD 5.5 million.

The implementation of the wastewater treatment plan is proposed through the Sewerage and Drainage Advancement Office. This is under the Director, Department of Public Works and Transport and Phnom Penh Capital City (PPCC). This Department is a part of the local government as well as of the Ministry of Public Works and Transport. This Ministry has a total of around 4500 employees. It executes projects totalling around USD 400–500 million per year. On 13 October 2016, a new dedicated sewerage department was formed. There are currently only 30 employees in this Department. One of their first tasks has been to prepare a sub-decree on the drainage and wastewater sector of Cambodia.

Given the current state of knowledge, expertise and preparedness of the Ministry of Public Works and Transport, it is likely to be very difficult for it to manage successfully even the first phase of the JICA-designed wastewater treatment plan (2035), even though it is at least some 6 to 7 years away. In addition, it is not clear how these projects will be funded. Currently the wastewater charges in Phnom Penh is 10% of the water tariff. Under the JICA plan (2035), the wastewater disposal tariff

---

[1] As per rule of thumb, if 100 L of water are consumed by a household, 75%–80% of it is discharged as wastewater.

needs to be increased to at least 70% of the water tariff to cover only the operation and maintenance cost of the wastewater management operations.

At a concessional loan interest cost of 5% a year, the first phase of USD 459 million will have an annual interest cost of USD 22.95 million. Along with the JICA wastewater management plan (2035), the total annual operation and maintenance costs for its operations will be USD 14.89 million. Thus, the total annual interest operation and maintenance costs for wastewater management is estimated to be USD 38 million. This is somewhat similar to the current annual water sales of PPWSA at USD 41 million. It is thus likely that the residents of Phnom Penh may have to pay as much for wastewater services as they pay for their drinking water. This additional tariff, plus the increase in water tariffs in the near future so that PPWSA can remain financially viable, may be tough to sell to the inhabitants of Phnom Penh, unless they are sensitised to the essentiality of these increased tariffs for their own well-being.

The consumers should also be made aware that when wastewater collection, treatment and proper disposal start, their combined water and wastewater monthly bills may be nearly double compared to their monthly water bills. This will become a reality as soon as the wastewater management systems come on stream. The consumers need to be sensitised to the idea of paying a higher water bill since the costs of providing good water services in the coming years will have to go up. PPWSA already has a very efficient operation and thus improving efficiencies further in the coming years will become an increasingly challenging task.

Unless the public understands of the need to pay substantially more for water and wastewater services, Phnom Penh is very likely to find it difficult to implement its wastewater management plan in the next decade. This will pose a big challenge to manage water quality. The general health and welfare of its residents will suffer because increasingly more untreated wastewater will have to be discharged in and around the city.

8.6 Listing on Cambodia Stock Exchange

In 2012, PPWSA was the first company to be listed in the Cambodian Stock Exchange. The company offered 15% of its shares during the initial public offering, of which 1.24% was reserved for its employees under the employees' stock option plan. There are ostensible benefits for public listing of a company, especially in terms of better corporate governance and improved transparency. PPWSA's listing was a testimony to its robust management and sound financial health. In retrospect, the listing has also helped to improve its transparency levels even further.

Notwithstanding the benefits received due to its public listing PPWSA may wish to explore the possibility of delisting its stock in the near future. This is because a strong economic and independent regulatory regime needs to be established before the listing of any urban water utility. The economic regulator for the urban water utility should have de jure as well as de facto power to adjust the tariffs to provide adequate financial returns to the utility. The regulator should further ensure that whenever tariffs are to be increased, the utility is operating as efficiently as possible, and the new tariffs are reasonable and affordable to a large majority of the inhabitants.

Economic regulation of urban water utilities in Cambodia is still in its infancy. The tariff decisions in the recent past have been primarily influenced exclusively by political considerations, and not by social and economic considerations to keep a good water utility functioning sustainably. These tariff decisions, unless rectified in the foreseeable future, will mean steady but progressively higher financial losses to the water utility. This is because the current marginal cost of producing water, at KHR 1600–1900/m^3, is higher than the current estimated average selling price of KHR 1200/m^3. The shareholders of PPWSA, naturally, are expecting PPWSA to deliver consistent profit growths. There is thus a fundamental dichotomy between the expectations of its shareholders and the socio-political objectives of the political leaders.

While PPWSA should strive to remain financially sound, its sole objective cannot be solely limited to making consistently higher and higher profits during the next five to ten years. Less than 20% of certain areas of Phnom Penh, like Chroy Changvar, Chbar Ampov, Prek Pnov and Dangkor now have access to water supplied by the utility (PPWSA 2017). Providing access to reliable supply of water to these areas, as well as additional new areas to which PPWSA may be instructed to provide water, will take significant additional expenses and investments. These costs are unlikely to increase PPWSA's annual profits consistently which the shareholders expect. Accordingly, PPWSA should consider the possibility of buying back its shares from the market. These shares then could be placed with sovereign wealth funds or pension funds. Sovereign wealth funds have bought shares in listed water utilities in England when some of the listed water utilities were delisted. Even the Government of Cambodia can consider buying back 15% of the listed PPWSA shares from the market. Funding for this purchase could be potentially supported by international donor agencies.

A delisted PPWSA will not have pressure from the minority shareholders to generate consistent profits over the short-to-medium terms. This will allow the utility to plan for the longer term. It may also then have a better opportunity to get more concessional loans from some of the multilateral donor agencies.

The Cambodian Government should further consider making PPWSA as the lead agency to implement the wastewater management plan for Phnom Penh. It has significantly more planning, management, and construction expertise than any other government departments. PPWSA is also more efficient than private sector companies of Cambodia in laying pipes and managing other technical and engineering aspects, both in terms of quality, costs and time. Under the present conditions, PPWSA cannot be responsible for implementing a wastewater management plan since this operation will most certainly remain unprofitable for a long time to come.

Shareholders or any publicly traded company will not want a business that is unlikely to make any profit over the short to medium-term. This is in spite of the fact that currently wastewater is the responsibility of the Ministry of Public Works and Transport and domestic water supply is the responsibility of PPWSA and MIH.

This institutional division of responsibilities is not a proper arrangement either technically or economically. At some stage Cambodia will have to merge domestic water supply and wastewater management services into one agency in order to have a streamlined and efficient system. Over the medium term, the Cambodian Government needs to consider an institutional consolidation by combining domestic water supply and wastewater and stormwater management functions under one entity. PPWSA is undoubtedly the best candidate by any consideration.

References

JICA (Japan International Cooperation Agency) (2016) The study on drainage and sewerage improvement project in Phnom Penh metropolitan area. Final report. https://openjicareport.jica.go.jp/pdf/12270294_01.pdf

Koemsoeun S, Handley E (2017) Exam pass rate increases. In: The Phnom Penh Post, Sept 11. https://www.phnompenhpost.com/national/exam-pass-rate-increases

PPWSA (Phnom Penh Water Supply Authority) (2017) Phnom Penh Water Supply Authority Third Master Plan, period 2016–2030, vol. 1 (Master Plan Report). Phnom Penh Water Supply Authority, Phnom Penh

World Bank (2019) Inflation, consumer prices (annual %): Cambodia. https://data.worldbank.org/indicator/FP.CPI.TOTL.ZG?locations=KH

Chapter 9
Lessons Learnt for Developing Countries

9.1 Introduction

By any criteria, both the operational and financial positions of the Phnom Penh Water Supply Authority, in 1993, were untenable. It has undergone major transformation during the 1993–2019 period. This can be best illustrated by noting the extent to which its business expanded during the 26-year period between 1993 and 2019. During this time, PPWSA's number of customers increased by 14.5 times (Fig. 9.1), volume of treated water by 14.2 times (Fig. 9.2) and number of metered connections by an incredible 115 times. Yet, its number of employees increased by less than two times during the same period.

What is remarkable is that this transformation process started in 1993 when Cambodia's per capita annual income was only USD 253 and it was amongst the poorest countries of the world (World Bank 2019). In addition, the country underwent long periods of political and institutional instabilities during 1970–1993, the extent of which very few countries in modern history had suffered. Considering the politically unstable and poor status of administrative, management and technical expertise available in the country in 1993, any realistic expectation that the water utility would turn itself into a functional, efficient and profitable institution that could provide clean water on a continuous basis, within only about one decade, must have been very close to zero.

Many urban water utilities of the developing world still face operational, financial and manpower challenges that PPWSA had faced in the 1990s. They can draw many lessons from the successful PPWSA experience. They can learn how the water utility was able to overcome the multifaceted challenges that it faced in the recent past. These challenges have been common to many of the cities of the developing world. However, unlike Phnom Penh, very few, if any, cities have been able to overcome them successfully.

This section discusses the key lessons that the urban water utilities of the developing world can learn from the PPWSA experience. It also outlines many of the key challenges that the water services sector is likely to face in Phnom Penh over

A. K. Biswas et al., *Phnom Penh Water Story*, Water Resources Development and Management, https://doi.org/10.1007/978-981-33-4065-7_9

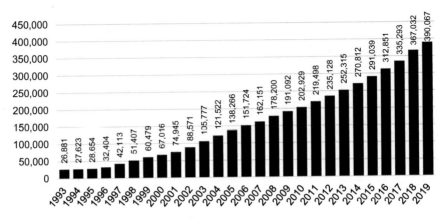

Fig. 9.1 PPWSA number of customers, 1993–2019. *Source* PPWSA

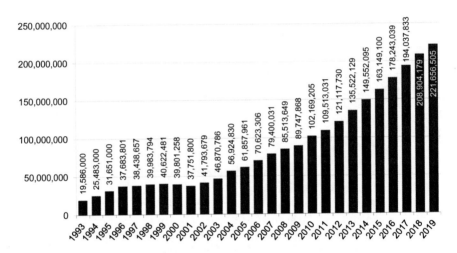

Fig. 9.2 Volume of treated water (cubic meters), 1993–2019. *Source* PPWSA

the coming short to medium-terms. The water service sector in this context includes drinking water supply, wastewater collection, treatment and stormwater disposal services.

Further, twenty residents of Phnom Penh were interviewed comprehensively to determine their views and experiences with different public service providers of that city, including water services. This section also summarises the key findings of the customer survey.

9.2 Key Learnings for Water Utilities in Developing Countries

9.2.1 Leadership

Nearly all the water utilities in cities of the developing world are suffering from physical, operational, financial, and institutional deficiencies. These cities have been facing tremendous challenges over the past several decades to transform their struggling water utilities into efficient and well-performing ones. Surprisingly, even though many of these cities have had access to adequate investments and operational funds, trained manpower, a thriving private sector to whom utilities could outsource many activities and good political stability, their performances have been well behind that of Phnom Penh.

Water utilities need strong, incorruptible and committed leaders who cannot only formulate an inspiring and exciting vision for the future, but also are capable of implementing that vision successfully. As a Japanese proverb correctly notes: "Vision without action is a daydream. Action without vision is a nightmare." Fortunately, PPWSA managed to transform the nightmare to a successfully implemented vision.

Fortunately for Phnom Penh, Ek Sonn Chan turned out to be a rare leader who could provide the vision, guidance, authority, control, commitment, inspiration, governance and direction to PPWSA during his stay as the Chief Executive from 1993 to 2012. As a capable, charismatic and strong leader, he attracted and retained key talents in the organisation, ranging from very senior to comparatively junior levels. He was able to motivate and retain the top and middle management, and successfully melded all its staff to become one dedicated team whose task became to take the institution to an uncharted but desirable and necessary territory. All staff members were proud to be members of a team which soon became well-known all over Cambodia as the most efficient, non-corrupt and functional public institution that provided good, consistent and affordable services to all the inhabitants, both the rich and the poor.

A strong leader does not wait for things to happen by themselves but works diligently and intelligently to make desired outcomes to become a reality. One good example is that Ek Sonn Chan successfully managed to raise funds when the institution was broken, and it was in dire need for additional significant financial resources. He was further able to convince the Ministry of Economy and Finance that it would be better and more desirable for PPWSA to negotiate loans directly with the bilateral and multilateral donors rather than the Ministry negotiating them on its behalf. Strictly by the then existing law, the Ministry was expected to negotiate each and every loan with international donors, even when the grants and loans were to be used exclusively by PPWSA. The direct negotiation of loans by PPWSA with donors reduced the time taken for approvals of loans. It also helped promptly implementation of the approved projects by PPWSA. In addition, PPWSA knew their internal conditions and constraints better than the Ministry. It also had significantly more knowledge and information on the relevant subjects than any Ministry could have.

Generally, many government departments and agencies lack internal competencies and knowledge as to how best to address the challenges they face. Currently the term capacity building is regularly used when it comes to improving the competences of the government institutions. However, there cannot be any effective and long-term capacity building in an institution unless there is a strong, competent, committed and dedicated leader at the highest level of the organisation. It is mostly neither realised nor appreciated that capacity building has to start from the very top of the organisation. The Head must understand and appreciate the manifold challenges the organisation faces, and likely to face, over the short, medium and long terms. Only then a strategy for skill and knowledge upgrading can be formulated and then implemented through further education and training. It further requires development of complementary human resource policies such as incentives and penalties for the employees in line with their actual performances on a regular basis.

In retrospect, Ek Sonn Chan and his senior management undertook all the requisite steps for capacity building, and then reinforced them with appropriate incentives and penalties that were commensurate with employees' performances. PPWSA took all appropriate steps to ensure that these tasks were conducted in a fair, prompt and transparent manner.

A starting point for revival of any urban water utility in developing world cities is to get a competent and passionate person to lead the organisation. The person should be committed and capable of bringing positive changes as quickly as possible. The Chief Executive should be a person who is able to lead from the front, and provide leadership for the institution by example, passion and dynamism. Ek Sonn Chan demonstrated his exceptional leadership qualities time and again during his entire tenure at the helm. He did it in the office and also in the field, along with his staff, to confront even the most powerful army generals who refused to pay the bills for the water consumed by the army, which happened to be Phnom Penh's largest water user.

There is also an urgent need in all developing countries to change the process of selection of the Chief Executive (CE) of an urban water utility. The CE of an urban water utility need not come from the bureaucratic rank as often has been the case in numerous cities of the developing world. Turning around a troubled and malfunctional water utility is not rocket science. It does not require very high education, or a high bureaucratic position. However, it must have committed, capable and passionate leaders like Ek Sonn Chan. Such leaders can turn a hopeless situation, as was the case in Phnom Penh in 1993, to an efficient water utility within a reasonable timeframe of say 8–10 years, which the nation, as well as the world as a whole, can admire.

9.2.2 Stability of Chief Executive Tenure and Culture of the Organisation

As much as an inspiring leader is essential at the helm of a water utility, the stability of the Chief Executive is an important requirement to ensure sound planning and then its subsequent implementation for a reasonable period of time. The practices and processes introduced by the leader can then become institutional culture. This will not happen even if a leader stays for only 1 to 3 years.

PPWSA has been fortunate to have a stable leadership at the top, first in Ek Sonn Chan who served as the Director General from 1993 to 2012, and then in Sim Sitha who has been the Director General from 2012 up to now.

The stability at the top of any water utility is one of the most essential requirements for its success. This also gives the Chief Executive time to change the entire culture of the organisation. Stability also ensures to retain key members of the management team. Every organisation has its own share of internal politics which is an undeniable reality of human fragility. Stability at the top helps in streamlining and maintaining proper working conditions between different departments of the institution and steers them towards a concept of teamwork for greater good of the organisation, rather than pursuit of individual glory and interest.

The long tenure of Ek Sonn Chan unquestionably helped to create a culture of performance, fairness and team spirit at PPWSA. The institution has continued to do well under its next Director General, Sim Sitha. PPWSA is thus an excellent testimony to the strong culture of teamwork and continued good performance.

The exemplary performance of PPWSA can thus be explained to a very significant extent by the stability at the top as well as a strong top management leadership team, provided by Chea Visoth, Ros Kimleang, Samreth Sovithiea and Long Naro and others. Each of these senior people have been with PPWSA for some 25 years. This has provided the organisation with continuity, stability and prevented unnecessary disruptions or detours in its culture and performance.

A competent and strong leader at the helm of an organisation for a reasonable length of time, often contributes to steady and continued permeation of good management and administrative practices lower down the line. This leads to development and maintenance of a strong culture of empathy, performance and team spirit in the organisation. Furthermore, an overwhelming percentage of PPWSA staff feels satisfied and is proud of their institution and receive significant professional and personal satisfaction from their work. This became steadily the case during the post-1996 period.

Any organisation has internal as well as external challenges. The leadership and culture of the organisation helps it to tide over most of its internal problems. The external constraints should not be an excuse for not solving internal problems. The sequencing of the solutions is important. Water utilities need to solve their internal problems, increase the effectiveness, credibility and impacts of the organisation, both internally and externally. Armed with a good reputation for problem solving, it can then successfully resolve the external challenges progressively over time. The leadership at PPWSA, quality and stability of its top management team, and their culture of performance and team spirit, enabled PPWSA to progressively improve their internal performance efficiencies without blaming the external constraints for their internal problems, as is often the case for many other similar utilities. This is an important lesson for other water utilities of the developing world to learn from, and to the extent possible, follow.

9.2.3 Autonomous Institution

In retrospect, granting of autonomy to PPWSA, the monopoly, under sub-decree number 52 in December 1996 (Annex I), was an important component for its success. With good leadership, its autonomous status allowed it to improve its modus operandi in many areas which it could not have done as a part of the normal Cambodian Government bureaucracy. One of the most significant benefits of becoming an autonomous institution was that its financial arrangements were no longer dependent on the administrative and bureaucratic requirements of the Phnom Penh Municipality, and also the inherent inertia that invariably exists which seldom allows for positive changes. This meant PPWSA was both accountable and responsible for its own operations and overall performance. Further, its accounts were no longer consolidated with those of the Phnom Penh Municipality. Conferment of autonomy ensured that its accounts and practices were transparent, and its responsibilities, and thus, its management had to take full responsibility for them. Earlier profits and losses of PPWSA were merged with the general accounts of Phnom Penh Municipality and there was little motivation or incentive for PPWSA to improve its financial performance as it had no control, or say, over the use of its income.

An autonomous structure further meant that on most issues PPWSA could take faster, better and more appropriate decisions that suited its objectives and interests the best. Unquestionably, its autonomous structure improved the top managements' morale and commitment to bringing the internal changes needed within the institution to make it as efficient and responsible as possible. For example, it had the authority to hire and fire employees based on its own objective assessments. The fact that it could dismiss errant employees following a transparent disciplinary hearing, improved very significantly the moral, performance and discipline within the organisation. The numbers of employees dismissed by PPWSA, for various valid reasons, from 2000 to 2017 is shown in Fig. 9.3.

Many Asian urban water utilities continue to be a part of the general municipality or another department of the local government. For example, the utility supplying water and wastewater services in Mumbai, India, is one of the largest in the world and perhaps is the second largest in Asia, after Tokyo. Responsibilities for the city's provision and management of water supply is with the Hydraulic Engineer Department. This is one of the oldest departments of the Municipal Corporation of Greater Mumbai. This was established under the MMC Act of 1888. In 2019, the population of Greater Mumbai was over 20 million people. It may be prudent and desirable to

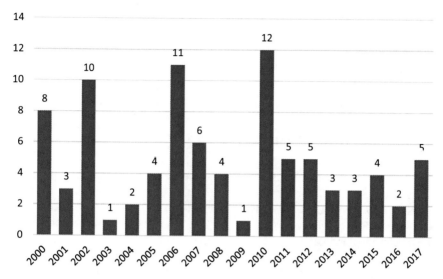

Fig. 9.3 Employees dismissed on disciplinary grounds, 2000–2017. *Source* PPWSA

have an autonomous urban water utility in cities like Mumbai. This may bring in more transparency, accountability, flexibility and efficiency in the operations of the Mumbai urban water supply system. If so, its inhabitants are likely to receive much improved water services from the city, compared to what they have received in recent decades.

Providing a reasonable and safe water supply to any megacity of more than ten million people is a difficult and complex task under even the best of circumstances. This is unlikely to be provided under a municipality that has also myriads of other responsibilities which may have higher political priorities. Water supply is seldom a priority consideration for managing most municipalities unless they are under prolonged droughts or heavy floods. Water supply and wastewater disposal of cities like Mumbai are complex and many. Their inhabitants, rarely, are satisfied with the services they receive. Making such water utilities autonomous institutions is likely to improve their performances as well as make them directly responsible for their own performance to its residents.

An autonomous institution is invariably capable of making decisions internally and thus its decision-making process is less complex. This also allows it to steadily build better institutional capacities and further enhances institutional memory of its learnings. An organisation that is dependent on decisions taken externally on its behalf, is seldom able to build and retain the skills necessary to navigate the challenges successfully and in a timely manner.

With urbanisation as a major megatrend in developing countries (Varis 2018), there will be increasing pressure on water utilities to provide continually better services for cities and towns. If these expectations of the public are to be met, water supply institutions in urban areas of developing countries need to be transformed into independent and autonomous entities. These utilities could be governed by a board of directors with greater operational and financial freedom and also higher flexibility. Operational and financial autonomies have been two important reasons for ensuring the success of PPWSA.

9.2.4 Non-revenue Water Reduction

Most of the Asian water utilities have non-revenue water around 25–40% and, it is estimated that 29 billion cubic meters of water is lost by the Asian urban water utilities every year (ADB 2010). At a production cost of water as USD $0.22/m^3$,[1] this volume of NRW is equivalent to an annual loss of USD 6.40 billion. In addition to these substantial regular annual financial losses, high non-revenue water invariably reduces water securities of the urban centres over the medium and long-term.

High non-revenue water not only reduces the financial sustainability of the urban water utilities but also puts additional burden on the consumers because of inefficient operation and poor maintenance practices that result in such high-water losses (Rouse 2020). The net result invariably is that consumers do not receive a proper and reliable water service.

Table 9.1 is an illustration that provides scenarios for two hypothetical utilities, A, which has NRW of 40% and B having NRW of 5%. Both water utilities are assumed to have a unit production cost of USD $0.20/m^3$. However, the unit breakeven price of water for Utility A is USD $0.33/m^3$ which is 50% more than Utility B. Thus, the customers of the inefficient Utility A have to pay 50% more compared to the consumers of the efficient water Utility B, assuming both the utilities break even on their costs. This is another good reason as to why all water utilities should endeavour to keep NRW as low as possible (Table 9.1).

It is absolutely essential for water utilities to save water by reducing non-revenue losses as much as possible. This will help the financial interests of both the utilities and their consumers. The steady process of NRW reduction in PPWSA as a priority

Table 9.1 Illustration of non-revenue water and its impact on break even costs

	Utility A (NRW at 40%)	Utility B (NRW at 5%)
Unit water production cost (USD/m^3)	0.20	0.20
Water produced (m^3)	100	100
Water sold for revenues (m^3)	60	95
Total water production cost (USD)	20	20
Implied break-even cost, based on water sold (USD/m^3)	0.33	0.21

---

[1] This was PPWSA's water production cost in 2016.

activity was initiated by Ek Sonn Chan. He adopted an eight-pronged strategy that can be used as a guide by any water utility of a developing country that wishes to reduce its NRW significantly over a reasonable time period.

1. **Effective leaks repair**—An approach of incentive and penalty was used for the employees of PPWSA to reduce the leaks in its system. A repair team was on standby, on a 24/7 basis, to repair any leak in the water network, at any time, on any day of the week.
2. **Updating customer base**—A customer database was updated with door-to-door surveys. A new and reliable database of 26,881 customers was established for the first time in 1994. It has been regularly and reliably updated since then.
3. **Metering of all service connections**—In 1994, four teams for meter installations were created. While in 1993, only 12.6% of the connections were metered, by the early 2000s, all the connections were metered. This has continued to be so ever since.
4. **Fight against illegal connections**—There were heavy penalties against staff members who may have colluded with the customers for facilitating illegal connections. An incentive scheme was introduced for finders and informers of illegal connections.
5. **Standardised design of last mile service connection**—Around 80% of the leakages generally occur in the last mile of service connections. Even today, all service connections are installed by PPWSA itself. This helps in safeguarding the last mile connection against any water pilferage as well as ensuring good workmanship to prevent leakages.
6. **Gradual replacement of the old supply network**—Old corroding pipelines were responsible for substantial physical loss of water. PPWSA started replacing all the old pipelines in 1995. Much of the old 288 km of pipeline network was replaced by 1999. Other pipes which may be susceptible to increased leakages should be regularly replaced with newer ones. In Singapore, its water utility replaces 2% of its network each year.
7. **District metering area programme**—In 2003, the district metering area programme was initiated. In 2004, two such pilot areas were created. By 2012, the total area covered by PPWSA was divided into 50 district metering areas.
8. **Introduction of internal service contracts**—A performance contract was signed between the management and water loss reduction team. It first started as a pilot in two zones. A year later, this was extended to all the eight DMAs. The performance contracts helped PPWSA to motivate employees to repair all leakages, inadvertent as well as wilful in the pipeline network.

9.2.5 Data Availability and Reporting

PPWSA has developed a very sound and transparent financial reporting. Due to its robust accounting practices and reliable and prompt financial reporting, PPWSA was the first company to be listed on the nascent Cambodian stock exchange. PPWSA has seen rotation of auditors each year because of legal requirements. This means that its accounts have been scrutinised regularly by different auditors for nearly a quarter of a century. These auditors have been mostly branches of international auditing companies.

An anecdote will illustrate PPWSA's continuing philosophy that if it is open with its data and information to outsiders, it can only be beneficial to the utility over the long-term. In 2006, the senior author of this book, Biswas, was requested by the Cambodian Prime Minister Hun Sen to visit PPWSA and see to what extent the institution has improved its supply of clean water to the Phnom Penh residents, and how its performance can be improved further.

The visit to PPWSA was organised at 2.00 pm. During this meeting, Ek Sonn Chan assembled all his senior staff. His only request to Prof. Biswas was that PPWSA would provide all the information Prof. Biswas would like to have in order to fulfil the Prime Minister's request. After his analysis and review of all the data Biswas had requested, Ek Sonn Chan made only one request. This was to give PPWSA a detailed and objective briefing on what they were not doing well and what else they should be doing but were not doing, and how its performance can be further improved. He further requested that Prof. Biswas should be as "blunt" in his criticisms and not "sugar-coat" them. According to Biswas, in five decades of advisory work all over the world, both in developed and developing countries, no other individual from any institution has made such a simple, straightforward and honest request!

By 6.00 pm of the visit on that day, Ek Sonn Chan requested his senior staff to provide all the information on PPWSA's performance that Biswas had requested. The realistic expectation of Prof. Biswas was that he would probably receive all the information requested by the end of the week, before he would return to Spain. Perhaps only limited information could be put together by next morning. He was completely flabbergasted that by 11.00 am the following day, all the data he had requested were given to him, neatly and properly organised! This is just only one example of the efficiency of the information system and data-driven nature of management of PPWSA.

During our 15 years of regular interactions with PPWSA, in terms of data accessibility, transparency and openness, it is unquestionably one of the leading water utilities from any city of the world, either developed or developing. Any data or information requested by one of the co-authors was promptly provided without any delay or hesitation.

Another good indicator of PPWSA's confidence in the quality of its data, and its firm belief that transparency and ready access to all relevant data to the people who may wish to use them, including external institutions and individual researchers, are essential requirements to develop and maintain a robust data management service. The three co-authors of this book have advised or assessed water issues in at least developed and developing 30 countries in all continents of the world. One will be indeed hard pressed to find even a single water utility anywhere in the world that is as free in providing data as PPWSA.

Nearly all water utilities of developing countries have weak financial reporting capacities. Good coordination between the commercial and the financial departments

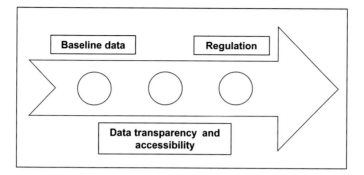

Fig. 9.4 Sequence of events for effective regulation of urban water sector

is absolutely necessary to generate proper data quality templates that can be used for effective management of water services business. PPWSA also excels on this count.

All urban water utilities should prepare detailed financial accounts on a regular basis. These should be audited by independent auditors and then made available on their respective websites for anyone interested to see and analyse them. Good and transparent data sharing with external stakeholders invariably contributes to improved generation of proper, reliable and usable internal reports. A good quality internal reporting will, in turn, help the water utility to improve its financial and overall operations, as well as its trustworthy reputation.

A sound financial and management reporting can provide the basis for comparing and benchmarking intra-country water utilities. Benchmarking of performance can be a precursor to the establishment of an effective economic regulation for the urban water sector. Economic regulation of urban water utilities, especially when they are unable to prepare good and reliable internal and external financial and management accounts in a timely manner, is unlikely to produce the expected results. The right sequence for bringing economic regulation is to build first the capacity of the water utilities to generate on a regular basis reliable baseline data, strong internal and external reporting and thereafter economic regulation of the water sector. This is shown in Fig. 9.4.

9.2.6 Standardisation of Treatment Plant Size

PPWSA has regularly expanded its water treatment plant capacities since 2007[2] in multiples unit sizes of 65 MLD. The Third Master Plan (2016–2030) has also proposed that all future water treatment plant capacities be planned in multiples of unit size of 65 MLD. The standardisation of water treatment capacities has been very useful in reducing manpower training, both in terms of cost and time, and also to ensure continued good operation and maintenance of these facilities.

---

[2]After Chroy Changvar plant expansions.

The standardisation has further reduced the spare part costs and also their storage requirements.

Many cities of developing countries have to regularly upgrade their urban water infrastructure. They can learn from PPWSA experience and adopt standardised designs for water treatment and wastewater treatment plants. The standardisation of design will also help to lower capital costs, reduce manpower training costs, and increase interchangeability of equipment and ideas across different locations.

9.2.7   Connections to the Poor

PPWSA has adopted the philosophy of providing clean water for the poor since 1999. This aims at providing clean water to the poor people of Phnom Penh who cannot afford to pay full costs. This has worked reasonably well. PPWSA water tariffs are 3–9 times cheaper than those of the private water providers in Phnom Penh. The poor customers do not have difficulty in paying for the water they consume. However, the poor households are often constrained to pay for the cost of upfront water connection. This is supported by a study done on Asian water utilities (McIntosh 2014). One of the main reasons for a lower percentage of poor people being connected to their respective urban water utilities in most developing countries has been the high costs for new connections. In order to make the water connection fee of KHR 430,000, in 2017, affordable for the poor, PPWSA provided three instalment options over 12, 17 or 22 months. Poor households can then select an option that suits their individual financial capacity the best.

The utility has also provided subsidies to the poor households, since 2005, with connection fees. The subsidy is provided in three slabs: 30%, 50% and 70% of the total connection fee. An additional category of 100% subsidy of connection fee was introduced in 2006.

In order to achieve the objective of serving the poor seamlessly and properly, a dedicated team identifies the poor households. The team then conducts proper poverty evaluations and scoring analyses. The team also communicates the "clean water for the poor" scheme to all likely poor households who may be entitled to these targeted subsidies, and then facilitates the filling up of the application forms on their behalf in situ.

"Clean water for the poor" policy has been supported by the World Bank, the Asian Development Bank and Mairie de Paris (PPWSA 2015). From 1999 to 2019, PPWSA had provided 32,613 connections, or 9.13% of total connections, to the poor through such schemes. The schemes have ensured that the poor households have regular access to 24/7 clean water that is drinkable straight from the tap, has sufficient pressure and is affordable.

PPWSA provides a detailed account in its annual reports under the section "Clean water for the poor." They provide granular details of the number of new connections in different areas of Phnom Penh, as well as information on customer engagement workshops with the poor households. The definition of poor has also evolved with

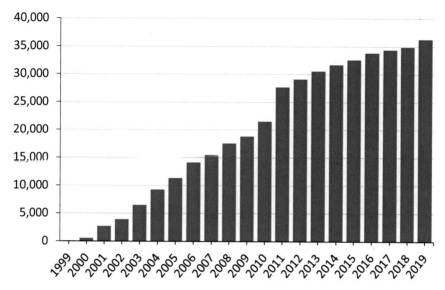

Fig. 9.5 Annual and cumulative water connections to poor households, 1999–2019. *Source* PPWSA

time. In 2017, a poor person was defined to be someone who earned less than KHR 65,000 per month or USD 158.[3] Figure 9.5 shows cumulative connections to poor households during the 1999–2019 period. The number of poor households connected has progressively increased from 101 in 1999 to 36,316 in 2019, a 360-fold increase over a period of 20 years.

Numerous water utilities in developing countries are politically forced to supply free water to the poor. However, all studies indicate that most of the world's poorest people often are paying some of the highest prices for domestic water supply (UNDP 2006), often ranging from 3–14 times of what their affluent counterparts are charged. In addition, the richer households have the convenience of getting their water piped into their households. In contrast, the poor have the inconvenience to buy water from water vendors whose qualities of supplied water are mostly of significantly worse quality compared to piped water the rich receives from water utilities.

It should be emphasised that PPWSA does not provide free water to the poor, but it does help them with water connections in an affordable manner which is in line with their financial means. Equally, they provide targeted subsidies to the poor. The water utilities of both developing and developed world can learn much from the experiences of Phnom Penh by emulating and adopting their strategies for serving the poor effectively, efficiently and properly.

---

[3] 1 USD = KHR 4100.

References

ADB (Asian Development Bank) (2010) The issues and challenges of reducing non-revenue water. Asian Development Bank, Philippines. https://www.adb.org/sites/default/files/publication/27473/reducing-nonrevenue-water.pdf

McIntosh A (2014) Urban water supply and sanitation in southeast Asia—a guide to good practice. Asian Development Bank, Philippines. https://think-asia.org/bitstream/handle/11540/2501/urban-water-supply-sanitation-southeast-asia.pdf?sequence=1

PPWSA (Phnom Penh Water Supply Authority) (2015) Water for all and customer education (2015) Annual report. Phnom Penh Water Supply Authority, Phnom Penh. https://www.ppwsa.com.kh/Administration/downloads/social/Annual_Report_on_Water_for_All_2015.pdf

Rouse M (2020) Can water professionals do more? Int J Water Resour Dev 36(2–3):325–337. https://doi.org/10.1080/07900627.2019.1685952

UNDP (United Nations Development Programme) (2006) Human development report 2006: beyond scarcity: power, poverty and the global water crisis. United Nations Development Programme, New York

Varis O (2018) Population megatrends and water management. In: Biswas AK, Tortajada C, Rohner P (eds) Assessing global water megatrends. Springer, Singapore, pp 41–59

World Bank (2019) Data: Cambodia. https://data.worldbank.org/country/cambodia

Chapter 10
A Management Tool Kit for Chief Executives

10.1 Introduction

Based on detailed analyses of PPWSA, PUB, National Water Agency of Singapore, and several other successful Asian water utilities carried out by the authors, a tool kit for possible consideration of the Chief Executives of water utilities in the cities of developing countries, both public and private, is proposed herewith. These guidelines may help the Chief Executives to manage their water utilities efficiently in a financially viable manner.

Before the guidelines are proposed, one a priori requirement is that the authorities responsible for appointing any Chief Executive select a person for the position who is competent, committed, and has the necessary background, expertise and knowledge to manage such an organisation. The person should be carefully headhunted, and be appointed for at least a five-year period, with clearly defined performance benchmarks which should be reached each year. The overall performance should be rigorously and objectively assessed every year in terms of actual achievements and meeting the agreed benchmarks. If the person performs the assigned tasks successfully, the incumbent's tenure may be extended for a maximum of another five years, with new performance benchmarks.

This is an important consideration for water utilities of all developing countries. Regrettably, much of the time, not enough attention is paid to select a competent person to run a water utility. Equally, the performance of the Chief Executive is seldom assessed objectively and comprehensively on a regular basis. Furthermore, in many countries, the Chief Executive is often appointed by either the principle of Buggins' turn where merit and performance do not count to any significant extent.

Consider the case of major Indian cities. The Chief Executive of water utilities are officers belonging to the cadre of Indian Administrative Service (IAS). Very few IAS officers who are appointed as the Heads of water utilities have any experience in running a utility of any type, or any reasonable background on water-related issues. Their average stay of these Chief Executives in office is around 18–24 months. For even a very brilliant and committed person, it is impossible to make any significant

© The Author(s), under exclusive license to Springer Nature Singapore Pte Ltd. 2021   123
A. K. Biswas et al., *Phnom Penh Water Story*, Water Resources Development and Management, https://doi.org/10.1007/978-981-33-4065-7_10

and perceptible improvements in water and wastewater services in less than two years, especially when it is considered that their knowledge of running any utility, or of water management, was rather limited when they started their work. Thus, after taking office, it takes them at least six months to understand the problems, constraints and complexities that the water utility may be facing. By the time they become somewhat knowledgeable, and start developing a plan on what should be done, and could be done given all the existing vested interests, their term of office is practically over. These IAS officers are then transferred to their next assignment on which they have to start again from scratch. This is not a good practice.

In addition, the deputy Chief Executive of any major Indian water utility is also an IAS officer, whose tenure equally suffers from similar short-term handicaps of around 18–24 months. Thus, very limited improvements can be seen during the tenure of any Chief Executive. Equally, they are seldom held accountable and responsible for their actions during their tenures, and no objective assessment is done of their performances in terms of improving the operational efficiencies of the utilities or their management practices.

Unfortunately, but not surprisingly, with this type of revolving management practices, the performance of all the Indian water utilities has left much to be desired over the past several decades. Compare this practice with the performance of PUB, Singapore's National Water Agency. In 1965, it was at a very similar level to those of major Indian cities like Delhi, Mumbai, Chennai or Kolkata. However, in little more than two decades, its performance improved very significantly, and PUB became one of the best water utilities of the world. During the same time, water and wastewater services of most Indian cities have either at best improved only marginally, or most of their performance indicators have declined significantly.

Some may claim that Indian cities do not have access to the same level of management capacities, or access to technology as is the case in Singapore. Regrettably, nothing is further than truth. India has produced, and continues to produce, excellent engineers and managers who are now running many of the most successful multinational companies that are in the *Fortune 500* list of major global companies. Thus, availability of management talent has never been a real issue. Equally, all technology Singapore now uses are easily available in the global market. Any country or utility has access to it.

An important difference between the water utilities of Singapore and India is that the Chief Executives of PUB are always selected on the basis of merit and competence. They hold office for two terms of three years, during which their performances are rigorously and regularly assessed. Subject to their performance, they run PUB for six years. Since the selection of the Chief Executive of PUB has always been rigorous, they have consistently run PUB competently for six years. This has meant PUB has progressively and substantially improved over the past decades. In contrast, Chief Executives of the various Indian water utilities are not carefully selected. In addition, they do not have more than 18–24 months to run them. Thus, at best, their improvements can be marginal.

If the Indian cities rigorously headhunt Chief Executives, and then give them reasonable time to stay in office, say at least five years, subject to good annual performance evaluations, probably this step alone will resolve 50–60% of the water problems the utilities are currently facing, probably within a decade.

10.2 Management Tool Kit: A Three Step Approach for Chief Executives

The Chief Executive of any water utility has three main objectives. The first objective is to make the water utility economically efficient. The second is to make the water utility socially equitable, and, the third is to make it environmentally sustainable. These three objectives often are in contradiction to each other. The situation becomes further complicated since these three objectives are often dependent on external factors, or institutions, on which they have limited or no control. However, even though the Chief Executive may not be able to directly change the external constraints, a capable person should be able to manage them in some fashion by improving the internal operations and functioning of the utility, which could, directly or indirectly lessen or even negate the impacts of the external factors or institutions.

Following ten recommendations can help to remove the internal inefficiencies of the utility. This will enhance its internal and external credibility and reputation.

1. Reactive versus proactive approach—A water utility has many day-to-day tasks that need to be immediately taken care of. For example, certain areas under its jurisdiction may not be receiving adequate quantity of water of appropriate quality or may not be getting piped water with sufficient pressure. These issues need immediate and regular attention from the management. They should be successfully resolved as soon as possible from the perspectives of the consumers.

There are often political pressures to act quickly on certain day-to-day operational issues. The danger for any Chief Executive is that the person might get too involved in solving issues that are created because of lack of proper planning or improper management. The utility head may become a reactive manager, responding to day-to-day events as they occur, instead of driving events towards agreed future goals. Otherwise, the Chief Executive will become busy every day, fighting short-term fires and thus will not have much time to reflect on long-term strategic issues which will invariably determine the future course or direction of the utility. Generally, there is seldom any sustained pressure from outside forces, such as political bosses, to think long-term. However, short-term issues even though they may be minor in nature, often receive immediate political or social pressures to be solved promptly. Thus, most Chief Executives of water utilities of developing country cities spend more time on solving short-term transient problems, and not enough time on preparing and implementing long-term strategic plans. If the Chief Executive has to make a positive impact on the future of the water utility, he/she must find enough time to consider the utility's long-term future while taking care of the day-to- day problems of the water utility.

2. Strategy and planning cell—It is necessary to create, or strengthen, a cell which should be responsible for strategic planning. This cell must be staffed with people who are capable of considering the institution's long-term needs and requirements from various angles. It may be called a Strategy and Planning Cell. It is likely that the water utility may already have such a planning cell. Depending on its capability and performance, it may have to be restructured and/or strengthened with new and capable staff members so that it can be transformed into a proper and functioning Strategy and Planning Cell.

Sometimes the difference between strategy and plan are not properly understood or appreciated. A plan is a programme, or a scheme, for definite purposes (Newton 2020). A plan may often be somewhat inflexible in nature. If plan "A" is not working, plan "B" may be adopted which could be very different compared to plan "A." Strategy, on the other hand, is a blueprint, a roadmap to accomplish a specific goal. It is flexible and open for adaptation whenever a change is warranted because of changed conditions (Newton 2020).

The main responsibility of the Strategy and Planning Cell is to envision developments over the long-term to achieve the goals and the objectives of the utility. It is not only responsible for making traditional long-term master plans for its future development but also responsible to build different scenarios for its future based on likely different conditions. In a volatile, uncertain, complex, ambiguous and rapidly changing world, the Chief Executive should be able to rely on various strategic options that should emanate from the Cell which could deal successfully with both the short- and long-term challenges.

The Cell should report directly to the Chief Executive. It should have competent and experienced team members who are well versed on the four domains already discussed earlier: physical, operational, financial and institutional. The role definition of the head of the Cell, or any other member of the team, should clearly stipulate that their tasks are related to strategies and planning only. They should not have any day-to-day operational or administrative responsibilities. It is easy for the team members to get fully involved in normal daily routine activities and thus their work on formulating long to medium-term strategies and plans become secondary instead of primary goals. The Chief Executive has to ensure that this does not happen.

The Cell should be responsible for preparation of master plans as well as scenario buildings for the water utility. Scenario building may include consideration of situations like what may happen if the industrial/commercial water demands in the future exceed projected demands, or what may be the financial impacts on the hydraulic design of the network if the overall demands increase by 50–100%, especially when a particular area under the utility's jurisdiction expands significantly compared to what was forecasted during the planning process.

The Cell should be responsible for insourcing all the required data, as well as all requisite data and documents from the external sources. They should decide what are the most critical datasets that are needed for the current and future management of the utility, and what are the best alternatives available to obtain such information from internal and/or external sources. In addition, the Unit should have the capacity to analyse the data that are being collected by the different departments of the utility, and ensure that their analyses, results and recommendations are regularly fed into the policymaking and management decision-making processes of the utility in a timely manner.

The Cell should receive data and information from all other departments of the utility on a regular basis. The Unit should ensure that the data that are being collected are easily accessible on a common platform, and also is in right formats that would ensure data stored are compatible and can be analysed easily. Such information should be available to all departments of the utility for their possible use whenever necessary.

The members of the Cell should be carefully selected. In addition to their knowledge, expertise and experience, they should be passionate about bringing positive changes and regular improvements in the functioning of the water utility. While competencies can be built, or even outsourced, it is hard to build passion or commitment, or outsource it. An organisation that has a strong institutional memory will become invariably strong. A strong and functional Cell can catalyse to ensure strong institutional memories, which should improve the performance of the utility over the long-term.

PPWSA is one of the very few water utilities of the developing world which has had such a functional strategic and planning unit for many years. Most other utilities, even when they have such a cell, the members often spend much of their time on short-term and routine activities. Consequently, the formulation of long-term strategies and plans becomes a secondary operation. Unfortunately, often strategic and long-term planning may not receive enough attention not only from the Unit but also from the Chief Executive who may be preoccupied with solving short-term discrete problems. Thus, the whole purpose of having such a Unit is mostly lost.

3. Human resource planning—The Chief Executive should undertake a comprehensive review of senior management positions and other employees who may be retiring during the next five years. A detailed hiring plan should be prepared following discussions with all departments in terms of their future manpower needs. The final strategy should be decided by the Human Resources Department and the Chief Executive. There should be a programme to identify the likely leaders of the next generation, and then appropriate mentorship arrangements should be made for them so that they can develop the knowledge and management expertise to take over senior management positions when the time comes.

It will be necessary to develop a yearly training programme as well. The programme content should meet the needs of the trainees to perform their work properly, and efficiently. To ensure that the employees take these training programmes seriously, and then they use this new knowledge to improve the qualities of their work, there should be a year-end assessment of employees who participated in each of the training programmes, both in terms of their performances as well as the relevance and impacts of the training programmes. The training programmes should then be modified and further improved to meet the future needs of the water utility. The employees who consistently perform well should be considered for accelerated promotions to the next levels. All these should be done in a transparent and objective manner.

4. Standard operating manuals—All the departments should consult the standard operating manuals for their key activities. These manuals should be updated and improved as and when necessary. For example, the water utility should have a standard operating manual on how to create a district metering area. The manuals should avoid repetition of earlier mistakes. Such manuals increase the institutional memories of the organisation and help to retain the knowledge of the earlier years.

5. Focus on operations and maintenance strategy—Water utilities need to plan for the requisite capital expenditures to increase and improve the access to water supply to their consumers, as well as steadily improve the qualities of water supplied. A capital expenditure plan for new asset creation, which does not follow operation and maintenance strategy, may likely reduce the extent and the qualities of the outputs from the beginning. One strategy to reduce the operations and maintenance costs can be to focus on good plans for asset creation. For example, consider a water utility that needs to install automatic meter readers on its pipeline network to replace the analogue meters. The new meters could malfunction if the rainwater enters them. The water utility could then focus on good workmanship for the installation of the new meters, along with a robust operation and maintenance strategy which could anticipate and solve any issues in terms of their potential malfunctions.

6. Focus on quality—The primary product sold by any water utility is potable water. This means water must be safe for direct human consumption. Water should not only be safe, but it is essential that consumers perceive it to be safe, a fact that is often disregarded by nearly all water utilities. Regrettably nearly all water utilities of developing and developed countries, in the past, and also at present, are more focused on the quantity of the water supplied than on its quality. Consumers of piped water all over the world are increasingly losing faith in the quality of tap water they receive. Currently, there are at least 3.5 billion people in the world who do not trust the quality of water they are receiving from their utilities. Thus, they boil water, or use bottled water, or use various treatment systems to improve water quality. This is in spite of the fact that nearly all utilities have water quality testing departments. The water quality testing department should be independent of the water production department. It is a very important issue, and thus the head of the water quality testing department should report directly to the Chief Executive.

A comprehensive water quality sampling should be formulated and implemented. It should take samples from not only water treatment plants, transmission and distribution water pipelines but also from randomly selected domestic connections. Careful considerations should be given to the number of water quality parameters that need to be tested. An annual water quality report should be published by the utility, and this should be made easily accessible to all its stakeholders. Furthermore, utilities should provide a web-based enquiry service on water quality for its consumers. Water quality information should be readily available to its customers. A customer should be readily seeing the results of the most recent water quality test done at a location that is closest to the customer's address. These actions are likely to enhance the trust of the consumers on the qualities of water they receive.

7. Behavioural scientists—Customers are at the core of any water business. For any water utility, the company should convince its consumers to lower their per capita water consumption, and also try to gain trust about safe quality of water supplied through the utility networks. The utility could consider hiring behavioural scientists, and people with good marketing backgrounds, who can understand the consumer behaviours, and their preferences and needs. Traditionally, water utilities are dominated by engineers, and they generally occupy key management roles. The Chief Executive needs to diversify the skills and knowledge of the top management team by inducting behavioural scientists into the team.

8. Strong internal reporting and transparency—A strong internal management reporting system is a prerequisite for any successful water utility. The goal should be to ensure that all the departments share information with other departments with least amount of friction. The Chief Executive should further ensure that the audited financial accounts of the utility, along with management discussions and decisions on operations, are available on its websites for all its customers and employees to see. This will contribute to improving the transparency of the utility and thus its credibility and reputation.

Strong internal control and reporting and consistent transparency on part of the water utility will make it ready for any kind of fair and economic regulations that may be imposed on it. Furthermore, good transparency is a kind of self-regulation, which will also improve the utility's credibility and standing, not only with its consumers but also with its other stakeholders like government institutions at different levels, policymakers and the media.

9. Face political heat exclusively—The very nature of water utility operations mean that the Chief Executives are exposed to all kinds of political pressures and interventions almost on a continuous basis. The Chief Executive should direct all its employees not to accept any calls from the politicians or the media. The employees should be instructed to politely redirect all calls from the politicians to the Chief Executive's office, no matter at whatever level of power and influence they may be. The Chief Executive should decide whether the politicians' requests or comments are appropriate and their potential implications on the performance of the utility. The Chief Executive should firmly but politely refuse the requests if they are not proper. This will help to improve the credibility, trust and the performance of the utility, both within and outside. The employees should be informed if they accept calls from political figures, and their actions are influenced by such interactions, they may be subjected to disciplinary actions by the management. The Chief Executive should field all such political calls. This way the employees can continue with their normal work without being pressured by politicians to do something which may be against the long-term interests of the utility.

10. Make presence felt on the field—The Chief Executive should not get accustomed to holding countless meetings with his top management to review the operations and the management practices. The person should get to know more of the business by being with staff on the ground and interact with its customers to get some idea about their perceptions on the functioning of the utility. Technology should be used to stay updated. However, the eventual feel of the business can come through regular interactions with employees at all levels of the organisation. This will help to improve the morale of the organisation as well.

10.3   Usability of Proposed Management Framework

The proposed management framework in the preceding sections is aimed at helping the heads of water utilities to change what is in their direct control. There are undeniable difficulties for water utility CEOs from internal as well as external sources. The problems of water utilities of developing countries are well documented, which include among other issues, absence of proper water tariff structure, make them financially viable, demoralised staff, inadequately trained staff, overstaffing, political interference, absence of good information system, and endemic corruption. At present, it appears highly unlikely that even 5% of Chief Executives of water utilities of the developing world, are appointed solely on the basis of merit, and also they may be in office for a reasonable period of time, say at least five years, so that they can make perceptible improvements in how the utilities are managed and operated. The management framework proposed in this section will work only if the heads of the utilities have reasonably long stay to understand and appreciate the problems and constraints they are facing and then take appropriate corrective measures to overcome them. They will have to steadily remove various internal inefficiencies and inconsistencies which would enable them to increase their credibility with different stakeholders such as customers, policymakers, and the media.

The sequencing of the solutions is thus critical. As any good spiritual guru may suggest, it is desirable first to change internally. Only then, external constraints are likely to appear less onerous and thus may become manageable. A good management framework aims at improving significantly the overall performance of the water utilities, and thus successfully meet the needs of the consumers to provide reliable water services. When the utility changes itself for the better, its customers are likely to be more willing to pay higher tariffs. This will go a long way to ensure its financial sustainability.

With steadily improving performance, and increasing satisfaction of the consumers, a utility is likely to receive better technical and financial support from potential bilateral and multilateral donors. As customer and donor satisfaction increases, a utility is likely to garner political support as well. Any transformational changes may appear difficult and unachievable at first. However, with good planning and management, and firm commitments and support from the top management, the water utilities should be able to transform themselves over time.

This is exactly what the Phnom Penh Water Supply Authority did over a short period: it transformed itself from a failed utility in 1993 to a most successful one in the developing world within only a short decade. Fortunately, its performance after more than a quarter century continues to be stellar. The water utilities of all developing countries can learn much from PPWSA. Probably the most important lesson policymakers can learn is that urban water management is not rocket science. Nearly all urban water utilities of cities in developing countries can emulate the example of Phnom Penh Water Supply Authority. They all can reach sustainable development goals in the area of water and wastewater management by 2030 by significantly improving their governance processes and practices.

Reference

Newton C (2020) Plan versus strategy: is there a difference? Centre for Management & Organization Effectiveness, Utah. https://cmoe.com/blog/a-plan-versus-a-strategy-is-there-a-difference/

Annex I

Royal Government of Cambodia

No. 52

Sub-Decree

On

Establishment of Phnom Penh Water Supply Authority (PPWSA)

Royal Government of Cambodia has seen

- The Constitution of the Kingdom of Cambodia
- The Royal Decree dated September 24, 1993 on the Appointments of the First and Second Prime Ministers
- The Royal Decree dated November 1, 1993 on the Establishment of the Royal Government of Cambodia
- Royal Kram No. 02 xx 94 dated July 20, 1994 on the Arrangement and Functions of the Cabinet of the Government of Cambodia
- Royal Decree No. 1094–83 dated October 24, 1994 and Royal Decree No. 1094–90 dated October 31, 1994 on the Re-arrangement of the Composition of the Royal Government of Cambodia
- Royal Kram No. 0196–08 dated January 24, 1996 on the Establishment of the Ministry of Interior
- Royal Kram No. 0196–18 dated January 24, 1996 on the Establishment of the Ministry of Economy and Finance
- Royal Kram No. 0696/03 dated June 17, 1996 on the General Guideline for the Establishment of the Public Enterprises
- Request of the Senior Minister and Minister of Economy and Finance
- Consent of the Cabinet of the Minister

© The Editor(s) (if applicable) and The Author(s), under exclusive license to Springer Nature Singapore Pte Ltd. 2021
A. K. Biswas et al. (eds.), *Phnom Penh Water Story*, Water Resources Development and Management, https://doi.org/10.1007/978-981-33-4065-7

Deciding.

Chapter 1: General Provision

Article 1: PPWSA is the public economic enterprise defined in article 26 of Royal Kram No. 0696/03 dated June 17, 1996 on the General Guideline on the Establishment of the Public Enterprises. PPWSA is the legal entity with financial and administrative autonomy. PPWSA is under the management of the Phnom Penh Municipality and has its headquarters in Phnom Penh.

PPWSA must implement all the articles of this sub-decree. For the necessary activity that is not stipulated in this sub-decree, the management of PPWSA must follow the Royal Kram No. 0696/03 dated June 17, 1996 on the General Guidelines for the Public Enterprises and other existing commercial laws and regulations.

Article 2: PPWSA has the mission to produce and distribute the water for general and public uses in the Phnom Penh Municipality.

To achieve this mission, PPWSA will allow making all necessary commercial and financial operations and fixed assets as following:

- Production and distribution of the water in the Phnom Penh Municipality and downtowns areas surrounding Phnom Penh.
- Expansion, increase and rehabilitation of the production and distribution system.
- Commercial operation of the existing water or produced water.

Article 3: The agents of PPWSA have the rights to access to the public places or private residents as needed for the installation and repairing of the water pipes, technical controlling of the production and distribution facilities and the recording of the quantities of the water usage.

The agents must dress in clear organisational uniforms and show their organisational identification cards during their missions.

Article 4: In its operation, PPWSA must follow the same principles and practices legally applied for the commercial entities. In this context, PPWSA must have the great autonomy to manage its activities for the efficiency and development.

Article 5: PPWSA must pay all taxes defined in the existing laws and regulations.

Article 6: The initial capital of PPWSA should be defined by the initial balance sheet jointly agreed by the Ministry of Economy and Finance, Phnom Penh Municipality and PPWSA.

When this sub-decree comes to effect, PPWSA has exclusive rights on behalf of the state on all recorded properties in its inventory list.

Article 7: The employees of PPWSA must be managed by a separate regulation approved by the Board of Directors.

In the transitional period, the Board of Directors of PPWSA will decide the measures for such management by its first meeting.

When the separate regulation is in place, all PPWSA employees will not be longer the civil servants of the state.

Chapter 2: Administrative Management

Article 8: PPWSA must be governed by the Board of Directors composing of the following members:

- Representative of the Phnom Penh Municipality—1 person
- Representative of the Ministry of Economy and Finance—1 person
- Representative of the Ministry of Interior—1 person
- Representative of the Ministry of Industry, Mines and Energy—1 person
- Representative of the Ministry of Public Work and Transport—1 person
- PPWSA's Employee Representative—1 person
- Director General of PPWSA—1 person

The representative of the Phnom Penh Municipality needs to serve as the president of the Board of Directors.

All the members of the Board of Directors must be appointed by the sub-decrees in accordance with the proposals of the member-related line ministries/agencies.

The PPWSA's Employee Representative must be elected by votes of the PPWSA's employee.

Article 9: All the members of the Board of Directors must be Cambodian nationality (Khmer), have the full civic rights and are not used to be the substantial punishment by the law and court. Moreover, the members of the Board of Directors must be selected from the active officials of the government who have at least five years of working experience or the non-civil-servant citizen with the age not more than 65 and with sufficient experience and high skills in economics or law. The tenure of the members of the Board of Directors is three years and this duration can be renewed.

Within their tenures, the members of the Board of Directors are not individually or commonly liable. In case a member commits the serious mistake, he/she will be removed any time by the sub-decree.

Article 10: The Board of Directors has the full rights to decide all matters of PPWSA and the rights to submit or approve all legal documents subjective to the existing laws relating to public enterprises.

The Board of Directors has the following duties:

- Approve the enterprise's plan as stipulated in article 19 of this sub-decree.
- Regularly evaluate the planned results and recommend the correcting measures.
- Approve the annual balance sheet and activities report of the enterprise.
- Approve, on the proposal of the Director General, the organisational chart, internal regulation of the enterprise, guideline, and salary scale of the employee in accordance to the existing laws and regulations.
- Provide the agreement on the contacts and treaties in which PPWSA is legally the party in accordance to the related existing laws and regulations.
- Approve sales, purchases and rents of all kinds of fixed assets or shareholders and approve the medium to long-term liabilities of PPWSA in accordance to the existing laws and regulations.

- Approve the establishment, opening and closing of the PPWSA's offices and agencies everywhere based on its judgment.

Article 11: The meeting of the Board of Directors can be called by the president or the proposal of at least four members. Such meeting should be held at least one per three months.

The acting president can be selected from the members of the Board of Directors.

The Board of Directors will discuss all the issues in its jurisdiction. The agenda of the meeting and the related documents must be informed to all members of the Board of Directors and the state observer in 10 days in advance of the meeting.

The meeting of the Board of Directors can only be held with the presences of at least 50% of the total number of the board members. In the case the quorum for such meeting is not met, the president needs to re-call for the meeting at least in next 15 days to discuss the same agenda. The latter meeting will not require the quorum.

All decisions of the Board of Directors can be considered valid only it is made with the majority of the approval of the members at the meeting. In the case the equal voice between the approval and disapproval exists, the vote of the president is used for the decision.

Article 12: All the decision of the Board of Directors must be recorded in a report that will be shared with the parental line ministries/agencies, the state observer and other related institutions within 15 days after the meeting as stipulated in article 25.

The report of the meeting must be approved officially by the next meeting of the Board of Directors and must be copied and kept in archive of the organisation in accordance to the administrative procedure. The original version must bear the signatures of the president of the meeting and a participating member.

Article 13: The Director General manages the daily operation of PPWSA. The tenure of the Director General is three years defined by the sub-decree of his/her parental line ministry/agency.

Article 14: The Board of Directors transfers all the necessary authority to the Director Generals for the daily operation of the enterprise in accordance to the existing law and guideline of the Board of Directors.

In this context, the Director General:

- Prepares all the documents for the approval of the Board of Directors and implement all decisions of the Board of Directors. The Director General will report regularly to the Board of Directors about the activities of PPWSA.
- Has the responsibility for the technical, administrative and financial management of PPWSA.
- Represents the PPWSA to all third parties in the civil, administrative and court letter in accordance to the existing laws and the decision of the Board of Directors.
- Has the rights to select, appoint, remove or fire and the rights to order the agents or employee of PPWSA in accordance to the existing laws and the decision of the Board of Directors.

- Has the rights to transfer part or whole of his responsibility to his subordinates for signing in accordance to the guideline of the Board of Directors.
- Implement all tasks assigned by the Board of Directors.

Article 15: Financial dividend for the Director General and the honorable allowance for the members of the Board of Directors must be defined by the related parental line ministries/agencies and the Ministry of Economy and Finance based on the proposal of the Board of Directors.

Article 16: The engagement in and implementation of the contracts of PPWSA must be in accordance to the existing laws and procedures.

Chapter 3: Financial Management

Article 17: PPWSA has its own account using Riel currency in accordance to the accounting regulations of Cambodia and following the Generally Accepted Accounting Standard (GAAS). The annual financial fiscal year starts from January 1 and finish in December 31.

Balances and the management of the accounts are verified by the Board of Directors before March 30 of the next fiscal year using financial reports submitted by the specialised accountant who is the financial inspector.

Balances and the management of the accounts as well as the financial reports of the financial inspectors must be submitted for the approval of the Ministry of Economy and Finance within 15 days from the date the Board of Directors reviews the documents. Those documents will be sent further to the parental line ministries/agencies for their comments in accordance to article 16 of the Royal Kram No. 0696/03 dated June 17, 1996 on the Guideline on the Establishment of Public Enterprises.

Article 18: The financial inspector must be appointed by the Board of Directors for two-year tenure. This period can be extended. The financial allowance for the financial inspector must be defined by the Board of Directors and is considered as the operational cost of PPWSA.

The financial inspector must produce the report providing the opinion on the transparency and regularity of the accounts in every fiscal year. The report reflects the realities of the financial and asset situation of PPWSA.

All the time, the financial inspector can check and control the accounts and can ask for all necessary documents and financial reports for such purpose. When the financial fraud or irregularities is found, the financial inspector must report immediately to the president of the Board of Directors.

The financial inspector can attend the Board's meeting based on the invitation of the president of the Board of Directors.

Article 19: Every year before October 1, the Board of Directors must provide the approval on the enterprise's plan based on the proposal of the Director General. After the approval of the Board of Directors, the proposal will be sent for the approval of the parental line ministries/agencies and the Ministry of Economy and Finance.

The enterprise's plan must consist of:

- Investment plan and its financing plan.
- Operational budget of the enterprise.
- Prices of the water and other services of PPWSA to ensure the revenue of the enterprise sufficient for covering its operational expenses (excluding the depreciations) and ensuring its financial balance.
- Criteria for measuring the economic and financial results of PPWSA.
- Subsidy by the state to PPWSA in order to support the loss in its public service delivery.

Article 20: The source of financial inflow of PPWSA can include:

- Principal capital provided by the state.
- Contribution or state subsidy or contribution from other public entities.
- Gifts or private contribution with the primary agreement of the Board of Directors.
- Liabilities.
- Revenues from the operation of the enterprise.
- Revenues from sales or rent of the current and fixed assets.

Article 21: Expenditures of PPWSA include:

- Current expenditures.
- Investment expenditures.
- Payment of the liabilities.
- Other expenses occurred to PPWSA in its operation.

Article 22: PPWSA can have the current account at the commercial bank and use this account based on its need.

Article 23: When there is the decision of the Minister of Economy and Finance, within and not more than five-year period, the net profit of PPWSA must be transferred to the reserve account for the strengthening of the financial ability of PPWSA.

Article 24: The medium to long-term liability transactions of PPWSA for its operation or investments must be agreed in advance by the Minister of Economy and Finance.

Chapter 4: Relationship with the Government

Article 25: Within 15 days after the approval of the Board of Directors, PPWSA must send to the minister in charge of the Council of Ministers and the ministers of the parental line ministries/agencies and the Minister of Economy and Finance the following documents:

- Meeting report of the Board of Directors.
- Enterprise's plan as stipulated in article 19 of this sub-decree.
- Operational report, balances and account management of PPWSA.
- Controlling report of the financial inspector.

Article 26: The Minister of Economy and Finance can suggest the government to appoint one state observer to PPWSA. That official can participate in the Board's meeting and can provide comments on all points in the agenda of the meeting but does not have the rights to vote. The state observer must perform his/her mission in accordance to the article 20 to 25 of the Royal Kram No. 0696/03 dated June 17, 1996 on the Guideline on the Establishment of Public Enterprises.

Article 27: PPWSA has the rights to disconnect the supply of the water or other services of its private customers and public institutions that fail to pay the water bills to PPWSA on time.

Chapter 5: Final Provision

Article 28: The unfair relationship of the PPWSA and its customers and all unlawful activities of the customers that damage the enterprise particularly the loss of the water, illegal connections etc., must be dealt by the existing laws and regulations.

Article 29: The co-ministers of the Council of Ministers, the co-ministers of the Ministry of Interior, the Minister of Economy and Finance, the Government Delegate for Phnom Penh Municipality must implement this sub-decree effectively from the date it is singed.

Phnom Penh, December 19, 1996

First Prime Minister—Second Prime Minister

Prince Norodom Ranarith—Hun Sen

[Signatures]

Proposed by H.E. Keat Chhon, Senior.

Minister, Minister of Economy and Finance

[Signature]

CC: Royal cabinet

- Secretariat of National Assembly
- Secretariat of Royal Palace
- Cabinet of First Prime Minister
- Cabinet of Second Prime Minister-As stipulated in Article 29
- Archive

Source: Biswas and Tortajada, 2009.

Annex II

Financial Support of Donor Agencies

#	Duration	Project name	Amount (USD)	Assistance Type
1	*UNDP/WB*			
	1993–1994	Improvement of customer survey and record keeping	2.80 M	Grant
2	*WB*			
	1998–2004	Urban water supply project	23.97 M	Loan
	2003–2008	Provincial and Peri-urban water and sanitation project	10.19 M	Loan
3	*ADB*			
	1997–2003	Phnom Penh water supply and drainage project	12.75 M	Loan
4	*JICA*			
	1993–1994	Master plan—target 2010	69.19 M	Grant
	1995–2003	Urgent rehabilitation works		
	2004–2006	Re-master plan—target 2020		
	2003–2006	Capacity building for water supply system in Cambodia		
	2010	Project for introduction of clean energy by solar electricity generation system	8.00 M	Grant
	2009–2013	Niroth Water Supply Project -Phase I (Part B)	32.23 M	Loan
5	*French Government*			
	1995–1998	Phum Prek reservoir restoration and filter backwash equip. replacement, and expansion of Chamcarmon treatment plant	14.77 M	Grant
6	*AIMF France*			
	2006–2007	Agreement for funding and assistance for project implementation (S&D of HDPE fitting and Valves for Sen Sok Community)	0.30 M	Grant
	2009	Supply and delivery of HDPE and DI pipes and fittings project for Phum Trapaing Achanh and Phum Ondoung	2.15 M	Grant
7	*Marie de Paris*			

(continued)

A. K. Biswas et al. (eds.), *Phnom Penh Water Story*, Water Resources Development and Management, https://doi.org/10.1007/978-981-33-4065-7

(continued)

#	Duration	Project name	Amount (USD)	Assistance Type
	2007–2010	Clean water for all project (household connection)	0.20 M	Grant
8	*AFD*			
	2003–2008	The extension of Phnom Penh suburb water supply system	6.00 M	Grant
	2009	Feasibility study of the Sourth branch of Phnom Penh transmission main	0.10 M	Grant
	2012	Extension of water supply system to the greater PP (GPPWSS)	0.30 M	Grant
	2007–2009	The extension of Chrouy Changwar water treatment plant (Phase 2)	16.15 M	Loan
	2009–2013	Niroth water supply project (Phase 1)	24.00 M	Loan
		Raw intake station and raw water transmission mains		Loan
	2013–2017	Niroth water supply project (Phase 2)	37.50 M	Loan

***Remarks:

- Grant: USD 103.81 M
- Loan: USD 157.79 M
- **Total: USD 261.60 M**

Source: PPWSA's Statistic, 2017.

Annex III

Sample Questionnaire Form

Interview

Introduction: Why this interview? A study is being conducted on water supply services in Phnom Penh. The objective is to recommend some of the required policy actions required in water and wastewater sector in Phnom Penh and Cambodia to serve the interests of the customers better. Also, the study is intended to share the experiences of Phnom Penh water sector with other water utilities in Asia.

Your data and name will not be shared with anyone (without your approval).

© The Editor(s) (if applicable) and The Author(s), under exclusive license to Springer   143
Nature Singapore Pte Ltd. 2021
A. K. Biswas et al. (eds.), *Phnom Penh Water Story*, Water Resources Development
and Management, https://doi.org/10.1007/978-981-33-4065-7

I. Basic Information

Age: ....

Address (Which Khan): ....

Email address: ....

Phone (optional): ....

Occupation: ....

Do you use smart phone?... ....

What is your mode of transportation?...................

What is your commute time to work?...

What is your monthly expenditure?.........................

II. Household Information

1. You stay at rented house or owned house?

     ....

2. How many people living in the house?

     ....

3. How old is your house?

     ....

4. Do you have sewerage connection at home?

....

5. If you are connected to a piped sewerage, do you have to pay for the fee? If yes, how much?
....

6. Do you know where your home sewerage ends up in Phnom Penh?
....

7. Do you have septic tank at home?
....

8. How often is the septic tank cleaned?
....

9. Do you do the septic tank clearance? If yes, what is the fee?
...*............

10. Is there good drainage around your house?
....

11. What cooking fuel you use at home?
....

12. Can you name the company that supplies you with:
   a. Water Supply: ....
   b. Sewerage: ......
   c. Electricity: ....
   d. Cooking fuel: ......

13. Do you have any complaints related to:
   a. Water Supply: ....
   b. Sewerage: ....
   c. Electricity: ....
   d. Cooking fuel: ....

14. Do you experience flooding around or at your house during the big rain? If yes, what is the dept of the flood? Please choose one measurement below:
   a. Ankle
   b. Shin
   c. Knee
   d. Thigh
   e. Waist

15. Duration of inundation
   a. < 30 mins
   b. 30 min-1 hour
   c. 2-3 hours
   d. 4-6 hours
   e. Almost half day

    f.   More than a day

16. What do you do with your solid waste?

....

17. Do you pay for your solid waste disposal?

....

III.   Utility using experience

1. What is your view on quality of water supply services? Now and say ten years back. What has changed?

...

2. Do you get 24*7 water supply?

....

3. Do you get good pressure in the pipe supply?

....

4. What is your monthly water bill?

....

5. Do you know how much water you consume in a month?

......

6. What is your view on quality of electricity supply services? Now and say ten years back. What has changed?

IV.   Improvement of utility quality in Phnom Penh City

1. If you had to change one thing in your life, what will that change be?

....

2. If you were the policy maker, what will you like to change the most?

....

3. Are you aware of any consumer forums to redress your complaints?

....

4. What can be done to improve the water supply in Phnom Penh?

....

5. What can be done to improve the sewerage system in Phnom Penh?

....

Annex IV

Unofficial Translation

Revised Water Tariffs

Kingdom of Cambodia

Nation Religion King

Ministry of Industry and Handicraft

No. 229 MIH/2017

Decision

On Modifying the Pricing for Clean Water

Treated and Supplied by Phnom Penh Water Supply Authority (PPWSA)

Senior Minister, Minister of Industry and Handicraft

- Having seen the Constitution of the Kingdom of Cambodia
- Having seen Royal Decree No. NS/RKT/0913/903, dated 24 September 2013, on the Appointment of the Royal Government of Cambodia
- Having seen Royal Decree No. NS/RKT/1213/1393, dated 21 December 2013, on the revision and addition of the composition of Royal Government of Cambodia
- Having seen Royal Krom No. 02/NS/94, dated 20 July 1994, promulgating the Law on the Organization and Functioning of the Council of Ministers
- Having seen Royal Krom No. NS/RKT/1213/018, dated 09 December 2013, promulgating the Law on the Establishment of Ministry of Industry and Handicraft
- Having seen Sub-decree No. 575/ANK/BK, dated 24 December 2013, promulgating the Law on the Organization and Functioning of Ministry of Industry and Handicraft

© The Editor(s) (if applicable) and The Author(s), under exclusive license to Springer Nature Singapore Pte Ltd. 2021
A. K. Biswas et al. (eds.), *Phnom Penh Water Story*, Water Resources Development and Management, https://doi.org/10.1007/978-981-33-4065-7

- Having seen Sub-decree No. 157/ANK/BK, dated 19 July 2016, on the Organization and Functioning of General Department of Potable Water Supply, Ministry of Industry and Handicraft
- Having seen PRAKAS No. 461 MIH/2014, dated 29 May 2014, of Ministry of Industry and Handicraft, on the Procedure for Issuing, Revising, Renewing, Suspending and Revoking the Water Supply Business Licenses
- Having seen PRAKAS No. 146 MIH/2017, dated 11 May 2017 of Ministry of Industry and Handicraft, on the implementation of the overall water tariff setting policy.
- Pursuant to necessary work requirements.

Decides

Article 1: Pricing for clean water treated and supplied by PPWSA is modified for the landlords who have rent their rooms to tenant workers and students, charging KHR700 for a cubic meter of clean water.

Article 2: The landlords as stipulated in Article 1 above shall resell clean water to tenant workers and students at a price of KHR800 per cubic meter.

Article 3: The revision and addition of water tariff scheme as stipulated in Article 1 and Article 2 above shall be implemented from 01 September 2017 onwards.

Article 4: Any provisions contrary to this decision shall be abrogated.

Article 5: Head of the Cabinet, Directors General, Inspector General, Directors of the Centers, PPWSA, Director of Phnom Penh and Kandal Provincial Departments of Industry and Handicraft, all the relevant entities and house owners or building owner shall effectively implement this decision from the date of its signature.

Phnom Penh, 29 August 2017 for and on behalf of Senior Minister of Industry and Handicraft Acting Minister

[Signature and Stamp]

SAT SAMY

CC:

- Office of the Council of Ministers
- Ministry of Interior
- Ministry of Economy and Finance
- Phnom Penh Municipality
- Kandal Provincial Administration

"To be informed"

- General Department of Potable Water Supply
- Phnom Penh Water Supply Authority
- Phnom Penh and Kandal Departments of Industry and Handicraft
- As in Article 5

"For implementation"

- Documentation and Archives

Index

Printed in the United States
by Baker & Taylor Publisher Services